Iowa
Birdlife

A Bur Oak Original

(Iowa Birdlife)

By Gladys Black

Foreword by Dean M. Roosa

Introduction by Carl Kurtz

PUBLISHED FOR THE NATURE CONSERVANCY

BY THE UNIVERSITY OF IOWA PRESS

University of Iowa Press, Iowa City 52242
Copyright © 1992 by the University of Iowa Press
All rights reserved
Printed in the United States of America

All articles which originally appeared in the *Des
Moines Register* are copyright © 1992, the Des
Moines Register and Tribune Company, and are
reprinted with permission. Permission to reprint
articles which appeared in the *Marion County
News* is gratefully acknowledged.

Printed on acid-free paper

96 95 94 93 92 C 5 4 3 2 1

96 95 94 93 92 P 5 4 3 2 1

Library of Congress
Cataloging-in-Publication Data
Black, Gladys.
    Iowa birdlife/by Gladys Black; foreword by
Dean M. Roosa; introduction by Carl Kurtz.
        p.   cm. — (A Bur oak original)
Includes index.
ISBN 0-87745-392-6 (cloth: acid-free paper),
ISBN 0-87745-393-4 (paper: acid-free paper)
    1. Birds—Iowa.   I. Nature Conservancy
(U.S.).   II. Title.   III. Series.
QL684.I6B58   1992   92-11196
598.29777—dc20   CIP

Contents

 # Foreword

I well remember meeting for the first time the author of this delightful book. It was at the banquet of the Iowa Ornithologists' Union, when she overheard me defending the habits of a sharp-shinned hawk which was hanging around someone's feeder. She agreed with me and later wrote to commend my stand. That was a long time ago, and she has since become perhaps the best-known ornithologist in the state, but she is quick to point out that she is "strictly an amateur, Bub, strictly an amateur."

Gladys Bowery Black was born within sight of the Red Rock bluffs and saw them daily on her way to school. These were a small girl's first "mountains." Her mother, Jerusha Bowery, knew all the local birds and taught Gladys, who at seven years could identify twenty-five species. After graduating from Pleasantville High School, she received a nursing degree from Mercy Hospital in Des Moines in 1930 and later a bachelor of science degree in public health nursing from the University of Minnesota. Her professional career began when she became a public health nurse in Clarke County, Iowa, working mainly with rural people. This was shortly after the public health and preventive medicine program was inaugurated by President Franklin Roosevelt. During this service she met Wayne Black, whom she later married. They moved to Georgia, where Wayne became administrative officer at Robins Air Force Base at Warner Robins and Gladys continued her public health career with the Public Health Service.

Although deeply involved with community affairs in Warner Robins, where she was named "Woman of the Year" in 1953, she reserved some time to polish and nurture her skill of identifying birds. She met two persons who were instrumental in the development of her interest in birding: Dr. David Johnson of Mercer University, Macon, Georgia (later of the University of Florida), who helped her establish a banding program, and Mrs. Tom Cater, a former student of Dr. Roberts', who aided greatly in identification.

When Wayne passed away in 1956, Gladys returned to Pleasantville to care for her mother. It was then that her impact on Iowa birds and birders began being felt as she immediately became active in the Iowa Ornithologists' Union, began a systematic collection of data on the avifauna of the Red Rock area, and began to write popular articles. She has recorded 305 species of birds at Lake Red Rock and has made bird observations every day for thirty-five years, except for five days when she was in the hospital!

A chance letter to Otto Knauth of the *Des Moines Register* about a kettle of broad-winged hawks resulted in Otto's request to print her letter in a column. This gave rise to a series of widely read articles in the *Register*, which eventually became the basis for her first book. Through her articles, talks, field trips, and many, many phone conversations, Gladys has done more than anyone else in Iowa to bring the science of ornithology to many laypersons. These articles are written in a popular vein, but they also occasionally asked that information on certain species be sent to her. The results of her surveys on evening grosbeak and snowy owl invasions, conducted through her columns, have added materially to our understanding of the extent of invasions by boreal species.

Aside from birding activities, Gladys has spent a great amount of time helping with other classrooms for schools in the country and conducting tours around Red Rock. She spends many afternoons with first and second graders, helping them gain insight into natural history. All youngsters should be so lucky!

Such devotion does not escape detection and recognition. In 1978, the U.S. Army Corps of Engineers presented Gladys with a certificate of appreciation for her conservation and education efforts around Lake Red Rock; the Iowa Academy of Science, also in 1978, presented her with an award of merit and elected her to the status of Fellow of the Iowa Academy in 1983; the Iowa Ornithologists' Union awarded her honorary membership at their 1977 meeting. In 1985 she was elected to the Iowa Women's Hall of Fame; in 1989, she was recognized by the state's governor for thirty-five years of volunteer work. But perhaps her most noteworthy award occurred on May 14, 1978, when Simpson College, Indianola, awarded her an honorary doctor of science degree for "splendid efforts as a humanitarian and a tireless defender of the environment which have reminded us of our obligation to posterity and to a world of beauty."

This book is a compilation of her articles that were printed in the *Des Moines Register* and those that appeared in the *Marion County News* from 1982 to 1992.

We in The Nature Conservancy are fortunate to obtain permission from Gladys and from the respective newspapers to reprint her articles. The profits of this printing will go to support Conservancy projects to preserve and protect the Iowa bird habitats described in this book. Your purchase of this book will provide you with a fund of knowledge about Iowa birds; it will also provide the Conservancy with a fund of another type, a fund to purchase and protect bits of the native Iowa that Gladys reveres—a native Iowa which is slipping away before our eyes. What we pro-

Dean Roosa banding a nestling red-tailed hawk held by Gladys Black, Robert's Creek Park, Red Rock, May 29, 1973. Photo by Herb Dorow.

tect in the next few years is what will remain for all time to remind those who follow what Iowa once was. I perceive this book as an ideal gift, a way to gain some insight into the lives of birds and a way to help young people gain a view of nature as seen through Gladys's eyes.

I have been fortunate in counting Gladys among my friends. Seldom do we have a chance to meet a truly dedicated person. Iowa has one in Gladys Black.

Dean M. Roosa

Introduction

Gladys Black's book *Birds of Iowa* was published nearly thirteen years ago. At that time we could not have foreseen its future popularity and demand. It went through two printings of 2,000 copies each, and the last copies were sold some years back. Gladys has continued to write about birds since 1979, and thus The Nature Conservancy felt it would be appropriate to republish the original articles along with a sizable number of new ones.

This book covers comprehensive information about life histories based on many years of observation and research for our specific geographic area. It is the kind of material that comes only with persistent dedication and covers about one-third of the common species that can be seen in Iowa on an average year. There are also accidental species, such as the magnificent frigatebird, and species which have infrequent invasions, such as the snowy owl and common redpoll.

This book has a dual purpose. First, it is intended to provide a readily accessible body of information about Iowa's birdlife for students and birders of every interest level. Second, monies generated from the sale of this book will be directed toward habitat acquisition through The Nature Conservancy's Iowa Chapter.

Numerous individuals have contributed to the production of this book. I would like to give the following acknowledgments. To the numerous photographers who have given me the use of their photographs. To Darrell Norman and Gary Hauser, who made the black-and-white prints under difficult conditions. To Dr. Jim Dinsmore, who gave his time and expertise to edit and check for minor technical errors. To TNC staff—Bev Allen, who shepherded the entire manuscript, Kim Chapman, whose community characterization work provided the habitat information, and all the other TNC staff members whose time, ideas, and energy made this edition a reality. And finally to Gladys Black, who gave us permission to use her material and has patiently put up with many delays in publishing the first and now the second edition.

Carl Kurtz

The Nature Conservancy

The Nature Conservancy is an international membership organization committed to the global preservation of natural diversity. Its mission is to find, protect, and maintain the best examples of communities, ecosystems, and endangered species in the natural world. Operating in the United States for over forty years and in Iowa since 1963, the Conservancy also has a Latin American Program that focuses on conservation of living resources in fourteen Latin American nations.

The Conservancy works by identifying lands that shelter the best examples of natural communities and species and determining what is truly rare and where it exists; protecting habitats and natural systems through acquisition by gift or purchase and assisting government and other conservation organizations in their land and natural diversity preservation efforts; managing more than 1.2 million acres nationwide (over 1,500 in Iowa) using both staff and volunteer land stewards and encouraging compatible use of the sanctuaries by researchers, students, and the public.

The Iowa Chapter was founded in 1963 by a number of interested private citizens and professional conservationists. Today members range from layperson to scientist, from casual bird-watcher to professional archaeologist, all of whom share a single goal: the preservation of natural environments and the species that live in them.

Business is conducted by chapter officers, a board of trustees (member-elected), various committees, and the director of the Iowa Field Office. The current director, Gary B. Reiners, assumed office on September 18, 1989.

Identification

The Iowa State Natural Heritage Inventory Program, a part of the state's Department of Natural Resources, is an ongoing inventory process that identifies rare natural elements and their locations within a particular state. Researchers use inventory techniques and assessment methods developed by The Nature Conservancy. The scientific information gathered by the inventory indicates the relative rarity of plant and animal species, aquatic and plant communities, and other significant ecological features.

The systematic inventory process also indicates which natural elements are currently protected and which are not. Consequently, the data can be useful in guiding development siting decisions, resource planning, and other conservation initiatives.

Protection

"Protection" is the term used by the Conservancy to denote the acquisition of property or interests in property that will protect the species of concern. A protection project is undertaken based upon a natural heritage program inventory that identifies a specific site sheltering critically threatened plant or animal species/communities or an acquisition strategy designed to enlarge an existing sanctuary.

The project is first reviewed by the Conservancy's senior management. If it clearly supports the Conservancy's mission, a purchase option is negotiated by Iowa Field Office staff. Upon approval, money for the purchase is made available from the Conservancy's national revolving loan fund, the Land Preservation Fund. The Iowa Field Office (or a special project committee) then raises funds to repay the national revolving fund so that the money can be reinvested in other protection projects.

Management and Stewardship

The Iowa Field Office manages or participates in the management of twenty-five preserves with a total of over 1,500 acres. It also has an active Natural Areas Registry Program with over five hundred properties enrolled. The registry program is a voluntary program that provides the private owners of lands with significant rare and endangered species information and assistance in the proper stewardship of their natural areas.

Acres acquired since 1963:	2,430
Acres managed:	1,500
Acres registered:	25,550
Location of projects:	34 Iowa counties
Membership:	5,000
Preserves with Conservancy management:	25

We need your help. For membership and volunteer information, please contact The Nature Conservancy, Iowa Chapter, 431 E. Locust, Suite 200, Des Moines, Iowa 50309, or phone (515) 244-5044.

Gary B. Reiners

Habitat Descriptions

During the past two centuries, Iowa's landscape has changed extensively and dramatically. Euro-American settlement and the resultant clearing of land for agriculture have had an enormous impact on the state's natural communities and their resident bird species.

As habitats were altered, some species such as the American robin were able to adapt to and exploit the new surroundings; others like the greater prairie-chicken could not. The reason for this is fairly straightforward. Birds are generally attracted to habitat on the basis of structure. They are typically concerned less with the plant species in a community than with the structure and shape of those plants. For this reason, grasshopper sparrows find exotic grasslands of smooth brome suitable replacements for native prairie. This is also why peregrine falcons are now finding that downtown metropolitan areas make good habitat. Consequently, many bird species occur today in habitats that did not exist before white settlement (e.g., agricultural fields, roofs of buildings, barn rafters, and urban areas).

For many rare species, suitable artificial habitat mimicking that which was once occupied in nature is no longer available. For example, no artificial substitute exists for rich (maple-basswood) forest. Habitat conversion, fragmentation, and other associated effects (both in Iowa and on winter ranges in Central and South America) have taken a heavy toll on populations of many forest-dwelling bird species.

Many prairie bird populations have also declined. Habitat fragmentation primarily through agricultural conversion has decimated our native prairie birds. Though thousands of acres of grassland have been planted across Iowa (mostly composed of exotic smooth brome), they do not fulfill the needs of many avian species. They are generally too small, too heavily grazed, too subject to heavy predation pressure, and too low in food diversity to possess diverse and established bird populations.

Protection of such natural habitats (e.g., rich forest, tallgrass prairie, and marsh), so crucial to the long-term health of many native bird species in Iowa, has been an ongoing priority of The Nature Conservancy for over forty years. For this reason, habitat descriptions in this chapter focus on the natural community types in the state prior to Euro-American settlement. It is within these natural habitats that the battle for the survival of many of Iowa's bird species will be won or lost.

Savanna

NATURAL COMMUNITIES: Dry loams savanna, dry sand savanna, dry limestone savanna.

DESCRIPTION: Savanna communities have scattered open-grown mature trees and patches of scrubby multiple-stemmed trees, shrubs, and open grassy areas; they are a mosaic of grassland and scattered trees. Trees often occur within the community on rises and slopes, at the margins of lakes and marshes, and as peninsulas of trees extending into prairies from adjacent forests. Savanna communities typically occur on level to rolling uplands of sandy glacial outwash, end moraines, or other areas of irregular, dissected terrain. Historically, the distribution of this habitat within the landscape was influenced most directly by the frequency and intensity of fire, which was very important in the origination and maintenance of savanna communities. Frequent, low-intensity fires created and maintained an open understory and increased grass and forb diversity but had little effect on canopy cover. Less frequent higher-intensity fires reduced canopy cover, promoted the growth of a dense sapling/shrub layer, and increased grass cover. Sand savannas historically experienced catastrophic fires on a regular (possibly every 80 to 120 years) basis. It has also been speculated that the canopies of rich savannas in Iowa were completely destroyed by periodic fires that swept off the large expanses of adjacent prairie, especially during drought years.

Plant species composition within savanna communities varies considerably and is tied closely with variation in topography and soils. Presumably, this is because of the effect topography and soils have on humidity, insolation, the pattern of disturbance (especially by fire), and availability of nutrients and moisture. The mature trees in a savanna stand typically are of the same age and size and are open grown with large, spreading lower branches close to the ground. The true dominants of the community are grasses rather than trees, as the grasses form a matrix in which the trees are conspicuous but incidental.

DISTRIBUTION IN IOWA: Historically, savannas were found throughout much of the state, particularly in the east and south.

PLANTS: In savanna communities, canopies are composed of a 5 to 60 percent cover by oaks—black oak (*Quercus velutina*), northern pin oak (*Q. ellipsoidalis*), white oak (*Q. alba*), bur oak (*Q. macrocarpa*), swamp white oak (*Q. bicolor*), pin oak (*Q. palustris*), and/or red oak (*Q. rubra*)—with subcanopies occupied by smaller immature oaks and other species. Common shrubs include American hazelnut (*Corylus americana*), poison ivy (*Toxicodendron radicans*), choke cherry (*Prunus virginiana*), and downy indigobush (*Amorpha canescens*). Herbaceous vegetation varies between that typical of a tallgrass prairie and that of a woodland, depending upon percent of canopy cover.

BIRDS: Bird species thought to be indicative of savanna include the eastern bluebird (*Sialia sialis*), American robin (*Turdus migratorius*), warbling vireo (*Vireo gilvus*), lark sparrow (*Chondestes grammacus*), orchard oriole (*Icterus spurius*), red-headed woodpecker (*Melanerpes erythrocephalus*), eastern kingbird (*Tyrannus tyrannus*), American goldfinch (*Carduelis tristis*), indigo bunting (*Passerina cyanea*), and American kestrel (*Falco sparverius*). Birds of shrub-dominated savanna include the brown thrasher (*Toxostoma rufum*), yellow-breasted chat (*Icteria virens*), white-eyed vireo (*Vireo griseus*), yellow warbler (*Dendroica petechia*), and Bewick's wren (*Thryomanes bewickii*). The demise of the bluebird over the past century has been the result of habitat loss through the destruction of savanna communities.

THREATS: Midwestern savannas are currently among the rarest natural communities in North America. There are virtually no intact examples left. The primary threat to savanna communities is fire suppression. With lack of fire in fire-susceptible areas there is increased canopy and subcanopy cover, invasion by fire-intolerant species, increased litter and fuel loads, and decreased herb-layer diversity. With loss of the natural fire regime, savannas have largely changed over time to forest communities. In the absence of fire, rich savanna succeeds to oak or oak-hardwood forest. Additional threats to extant savannas include excessive grazing (which may promote invasion by exotic species and eliminate successful estab-

lishment of young trees), hydrologic changes (which affect wetlands within savanna communities), agriculture, sand mining, and residential and industrial development.

Tallgrass Prairie

NATURAL COMMUNITIES: Dry loams prairie, dry sand prairie, dry gravel prairie, dry sandstone prairie, dry limestone prairie, dry quartzite prairie, mesic loams prairie, mesic sand prairie.

DESCRIPTION: Tallgrass prairies are grasslands with little or no tree cover. Prairies frequently grade into oak savanna and denser oak forest, although there may be an abrupt boundary between grassland and forest, particularly at rivers or at topographic breaks. On moist soils, prairies grade into marshlands dominated by sedges. Dry and mesic prairies formed over a wide variety of substrates ranging from sands deposited by glacial outwash, limestone ridges and bluffs, gravel slopes of eskers, kames, river terraces and glacial till, lacustrine deposits in glacial lake beds, alluvial deposits, and residual or loessial soils over dolomitic bedrock. Soils in this community are moderately moist to very droughty. Depending on slope orientation, standing water may be present, but only following heavy rains and snow melt.

DISTRIBUTION IN IOWA: Statewide.

PLANTS: Dominant plant species vary between locations as soil moisture, exposure, and substrate type fluctuate but typically include little bluestem (*Schizachyrium scoparium*), porcupine needlegrass (*Stipa spartea*), big bluestem (*Andropogon gerardii*), Indian grass (*Sorghastrum nutans*), switch grass (*Pa-*

nicum virgatum), and side-oats gramma (*Bouteloua curtipendula*). In mesic sites, gray-head prairie coneflower (*Ratibida pinnata*), cream wild-indigo (*Baptisia bracteata*), compass plant (*Silphium laciniatum*), shooting-star (*Dodecatheon meadia*), rattlesnake-master (*Eryngium yuccifolium*), and cattail gay-feather (*Liatris pycnostachya*) are common. Drier sites are commonly inhabited by downy indigobush (*Amorpha canescens*), American pasque flower (*Pulsatilla patens*), narrow-leaved puccoon (*Lithospermum incisum*), Pacific wormwood (*Artemisia campestris*), western silvery aster (*Aster sericeus*), and thoroughwort brickellbush (*Brickellia eupatorioides*). Tallgrass prairies have a high content of nitrogen-fixing legumes, which contribute to the high biomass production characteristic of the community.

BIRDS: Bird species frequently associated with tallgrass prairie include upland sandpiper (*Bartramia longicauda*), bobolink (*Dolichonyx oryzivorus*), meadowlark (*Sturnella* spp.), dickcissel (*Spiza americana*), savanna sparrow (*Passerculus sandwichensis*), and Henslow's sparrow (*Ammodramus henslowii*). In the Loess Hill prairies of western Iowa, the very rare lark bunting (*Calamospiza melanocorys*) is sometimes observed.

THREATS: Fires are important in maintaining prairies. Mesic tallgrass prairies are very stable when burned frequently. It appears that climate, substrate conditions, and fire were important in the origination of tallgrass prairie, while substrate, fire, and, to a lesser extent, grazing animals are important in maintaining the community. With fire suppression and elimination of large herds of native herbivores, tallgrass prairie succeeds to forest, woodland, and scrub communities. Historically, most of Iowa was vegetated by tallgrass prairie. Today, less than 1 percent remains, having been destroyed by cultivation, overgrazing, and development.

Rich Forest

NATURAL COMMUNITIES: Mesic loams forest, mesic bottomland forest, mesic limestone forest.

DESCRIPTION: Rich forests are closed-canopy maple or maple-basswood forests with ground layers composed primarily of species of low stature (less than 45 cm tall), including high densities (as many as 20,000 individuals per acre) of very shade-tolerant maple seedlings. Canopy coverage is essentially continuous by late spring so that very little light reaches the forest floor. As a consequence, the conspicuous herbs in the community are spring ephemerals and other spring-blooming species.

Frequent years of high sugar maple seed production saturate all favorable spots on the forest floor with seeds. Sugar maple seedlings are highly shade-tolerant and may persist in a suppressed state for decades. When an opening occurs in the canopy, an established maple seedling is usually present to replace the one that has been removed. In this way, sugar maples come to occupy a greater and greater share of the canopy of any forest in which they occur. Once established in the canopy, sugar maples persist because of their great longevity. It is the most fire-sensitive tree species in Iowa and is easily killed even by ground fires. Closed-canopy rich forests, however, are highly resistant to burning because of the high moisture content of the litter in the winter and spring and the frequent absence of inflammable litter in the summer.

DISTRIBUTION IN IOWA: Primarily in northeast and east-central Iowa, but also to some extent in the protected areas within the central part of the state.

PLANTS: This community may be dominated by sugar maple (*Acer saccharum*) or a mixture of sugar maple and American basswood (*Tilia americana*). Often red oak (*Quercus rubra*) may codominate. American elm (*Ulmus americana*) often was a dominant species in stands of this community before the occurrence of Dutch elm disease in North America. Spring ephemerals include large-flowered bellwort (*Uvularia grandiflora*), bloodroot (*Sanguinaria canadensis*), blue cohosh (*Caulophyllum thalictroides*), and Dutchman's breeches (*Dicentra cucullaria*). Shrubs and woody vines contribute little to forest structure. Fungi, however, may be numerous, in both number of species and individuals. The even moisture supply and richness of the soils are favorable for various fleshy fungi, especially in the group Ascomycetes. Cup fungi (like *Urnula craterium*, *Peziza badia*, and *Sarcoscypha coccinea*) are common in the early spring, and several species of morels (*Morchella* spp.) occur in the community.

BIRDS: The birds of rich forests are generally insect eaters. Characteristic species include cerulean warbler (*Dendroica cerulea*), red-eyed vireo (*Vireo olivaceus*), Acadian flycatcher (*Empidonax virescens*), hairy woodpecker (*Picoides villosus*), ovenbird (*Seiurus aurocapillus*), red-bellied woodpecker (*Melanerpes carolinus*), pileated woodpecker (*Dryocopus pileatus*), great crested flycatcher (*Myiarchus crinitus*), and American redstart (*Setophaga ruticilla*).

THREATS: The major impact by humans on rich forests has been through clearing land for cropland. Where stands have escaped clear-cutting and plowing, selective logging of oaks has tended to produce forests with greater proportions of sugar maple, basswood, eastern hop hornbeam (*Ostrya virginiana*), and other shade-tolerant species relative to oaks. As a result, there is considerable acreage of maple forest on lands originally dominated by oaks. Rich forests have little value as pasture because browse production by shrubs and grasses is very low. In stands that are converted to pasture there is a significant alteration of the understory, favoring those species able to tolerate or thrive with physical disruption and compaction—e.g., Virginia waterleaf (*Hydrophyllum virginianum*), mayapple (*Podophyllum peltatum*), raspberry (*Rubus* spp.), and gooseberry (*Ribes* spp.).

Dry Oak Woodland

NATURAL COMMUNITIES: Dry loams forest, dry limestone forest, dry sand forest, dry sandstone forest.

DESCRIPTION: Dry oak woodlands are closed-canopy oak or oak-hickory forests (55 to 100 percent cover) that occur on thin soils over bedrock outcrops on ridges, glacial outwash, ridges on sandy glacial lake plains, sand dunes, coarse-textured end and ground moraines, gravel river terraces, and loess. The sparse branching of the oaks enables considerable light to penetrate the canopy even when the canopy is nearly continuous. Shrubs grow thickly in the understory, and there are more summer-blooming herbs and fewer spring-blooming herbs than in rich (maple-basswood) forests. This community may be composed of relatively pure stands of one of the dominant oak species or it may be a mixture of these species. In the Driftless Area of northeast Iowa, white pine (*Pinus strobus*) frequently occurs in association with dry oak woodland.

Windthrow and oak wilt fungus cause local openings in the canopy in stands of this community. Relative to rich forests, dry oak woodlands are successionally unstable. Dominance by any one oak species is likely to last only one generation as other species with greater shade tolerance will replace it. This successional replacement is influenced by the topography of the site and is most rapid on flat lands, where the light on the forest floor is controlled entirely by the tree canopy. A solid canopy rarely forms on steep hillsides, so direct penetration of light from the side is always possible. Under such conditions even the most intolerant species will occasionally receive enough light for some successful reproduction and will be a continuing component of the community.

Topography is also an important influence on moisture conditions. A southwest-facing slope is considerably hotter and drier than a northeast-facing slope. Due to the extreme conditions on the southwest-facing hillsides, buildup of soil organic matter and the consequent retention of soil water are greatly retarded. Xeric species like yellow (*Quercus muhlenbergii*), bur (*Q. macrocarpa*), and northern pin (*Q. ellipsoidalis*) oak are dominant for long periods.

The understory species in white pine/oak relict forests in the Driftless Area of northeast Iowa are not uniform between stands. This is probably because of the small sizes of the stands and because the stands are isolated from one another. The fossil pollen record indicates that these relicts have existed in the Driftless Area for at least 12,000 years. They may have been present for much longer periods, most of that time isolated from the major areas of pine forest farther to the north. In Iowa, white pine/oak woodlands are slowly giving way to other community types.

DISTRIBUTION IN IOWA: Statewide but perhaps most abundant in eastern and southern Iowa.

PLANTS: Dominant tree species include the red (*Quercus rubra*), black (*Q. velutina*), white (*Q. alba*), and bur (*Q. macrocarpa*) oaks. Understory shrubs include choke cherry (*Prunus virginiana*), gooseberry (*Ribes* spp.), stiff dogwood (*Cornus foemina*), American hazelnut (*Corylus americana*), and raspberries (*Rubus* spp.). Forbs include tall hairy groovebur (*Agrimonia gryposepala*), large yellow lady's-slipper (*Cypripedium pubescens*), pale leaf sunflower (*Helianthus strumosus*), tall blue lettuce (*Lactuca biennis*), old field cinquefoil (*Potentilla simplex*), and false solomon's seal (*Smilacina racemosa* and *S. stellata*). Common vines include Virginia creeper (*Parthenocissus quinquefolia*) and smooth herbaceous greenbrier (*Smilax herbacea*).

BIRDS: Common birds of dry oak woodlands include scarlet tanager (*Piranga olivacea*), black-capped chickadee (*Parus atricapillus*), downy woodpecker (*Picoides pubescens*), rose-breasted grosbeak (*Pheucticus ludovicianus*), northern cardinal (*Cardinalis cardinalis*), blue jay (*Cya-*

nocitta cristata), northern oriole (*Icterus galbula*), red-eyed vireo (*Vireo olivaceus*), rufous-sided towhee (*Pipilo erythrophthalmus*), and wild turkey (*Meleagris gallopavo*).

THREATS: In much of Iowa, woodlots composed of oak forest are frequently grazed or selectively cut. Excessive grazing results in the destruction of tree seedlings, followed by the destruction of the shrub layer and taller members of the herb layer. Native grasses and sedges with growing points close to the ground increase in density, and exotic weeds of similar habitats, such as common brownseed dandelion (*Taraxacum officinale*), chickweed (*Stellaria* spp.), and Kentucky bluegrass (*Poa pratensis*), invade. Soil compaction and physical damage to tree trunks cause trees to die at an accelerated rate without chance for replacement by seedlings. Consequently, canopy cover decreases over time. The final result is the gradual change from forest to a savannalike community with a few trees scattered throughout a bluegrass pasture. Selective logging operations hasten successional changes and may result in a complete conversion from oak forest to maple-basswood or more usually to maple–slippery elm forests if a small amount of grazing has occurred. Where more than 50 percent of the canopy has been removed, there is a resurgence of intolerant herbs, shrubs, and tree seedlings that may have been present in the groundlayer, frequently resulting in impenetrable tangles. The oaks are all capable of sending out stumpsprouts, and second-growth forests are recognizable by the prevalence of trees with two or three or more equal-sized trunks that originated as sprouts.

In the absence of frequent fire, savanna communities become dry oak forests. Dry oak forests may also have been maintained in the past by some incidence of fire. Since the onset of fire suppression, mesic tree species have been invading many stands of oak forest in Iowa, and they are succeeding to rich forest.

Marsh

NATURAL COMMUNITIES: Stream marsh, sand basin marsh, basin marsh.

DESCRIPTION: Marsh communities consist of shallow basin wetlands formed in association with low floodplain terraces of streams and rivers, glacial depressions (i.e., potholes), river valleys, and lake plains. Marshes can also form within areas of slow-moving water along streams and rivers. Marshes possess standing water during much of the year, with depths rarely exceeding 6 feet. The marsh basin may become dry in mid to late summer or during drought. The waters are neutral to alkaline. Seasonal flooding occurs in the winter and spring or during heavy rains.

Marshes are wetland communities dominated by water-loving herbaceous plants. The vegetation in lowland freshwater marshes varies highly in response to changes in water depth and other forces. The vegetation includes submerged, emergent, and floating-leaved aquatic plants. Although the marsh community may include areas of open water, the majority of the community is occupied by vegetation. Large portions may be dominated by a single species. Shrubs may also be present. The important members of this community possess bulky underground parts, spread vegetatively by means of rhizomes, and have a marked development of internal passageways for gas transfer. The vigorous rhizome development of many species tends to produce a strongly colonial type of arrangement. Plant composition is a mixture of forbs and varies with respect to water hardness, clarity, soil type, and geographic location.

Marsh communities are influenced by the frequency and duration of flooding, depth of water, and animal (i.e., muskrat) disturbance. Water movement replenishes fresh water and circulates nutrients and organic debris through the system.

DISTRIBUTION IN IOWA: Statewide but most abundant in north-central Iowa in association with the Des Moines Lobe of the Wisconsinan glaciation.

PLANTS: Typical dominant species include cattails (*Typha angustifolia* and *T. latifolia*), sedges (*Scirpus* spp. and *Carex* spp.), broadleaf arrowhead (*Sagittaria latifolia*), common reed (*Phragmites australis*), smartweed (*Polygonum* spp.), large bur reed (*Sparganium eurycarpum*), duckweed (*Lemna* spp.), and greater bladderwort (*Utricularia vulgaris*).

BIRDS: Marshes provide a favored habitat for a large variety of animals. Common bird species include the mallard (*Anas platyrhynchos*), blue-winged teal (*A. discors*), northern shoveler (*A. clypeata*), American coot (*Fulica americana*), Canada goose (*Branta canadensis*), American bittern (*Botaurus lentiginosus*), least bittern (*Ixobrychus exilis*), several rail species, red-winged blackbird (*Agelaius phoeniceus*), yellow-headed blackbird (*Xanthocephalus xanthocephalus*), marsh wren (*Cistothorus palustris*), pied-billed grebe (*Podilymbus podiceps*), and swamp sparrow (*Melospiza georgiana*).

THREATS: Most marshes in Iowa have been drained and converted to cropland or destroyed by siltation. For those still present, siltation is a continuing threat.

Floodplain Forest

NATURAL COMMUNITY: Wet bottomland forest.

DESCRIPTION: Floodplain forests occur in moist to wet lowlands along rivers and streams and are characterized by a canopy cover of 50 percent or greater, a variable understory of smaller trees, shrubs and vines, and a ground cover of shade-tolerant herbs, lichens, and mosses. The tree canopy is closed, dense, and tall, ranging from 90 to 140 feet. The understory reaches heights of 25 feet or more and is also well developed.

Frequent and sometimes prolonged flooding (especially during spring) often gives way to drought conditions after waters recede. Flooding has a great effect upon the biota of this community. The conditions for germination are poor because of flooding. Many of the tree species have a multistemmed nature; it is thought that this characteristic is the result of damage to the trunk bases from ice floes and other flood-carried debris during the spring period of submergence. An unusually high content of woody vines (lianas) is present in the lowland forests. This is in contrast to the relatively insignificant role of shrubs in the community. Substantial changes take place in the composition of the dominant groundlayer community from year to year; this is in response to the differing abilities of the plant species to recover from the effects of the flooding. Due to flooding, full development of the groundlayer does not occur until mid-August.

DISTRIBUTION IN IOWA: Statewide along major rivers and large streams. Only along the Mississippi and Missouri rivers and large tributaries did expanses of this community exist.

PLANTS: Dominant tree species include silver maple (*Acer saccharinum*), American elm (*Ulmus americana*), green ash (*Fraxinus pennsylvanica*), black willow (*Salix nigra*), peach-leaved willow (*S. amygdaloides*), eastern cottonwood (*Populus deltoides*), common hackberry (*Celtis occidentalis*), sycamore (*Platanus occidentalis*), swamp white oak (*Quercus bicolor*), pin oak (*Q. palustris*), big shellbark hickory (*Carya laciniosa*), and river birch (*Betula nigra*). Herbaceous layers are dominated by nettles.

BIRDS: A rich avifauna is present in these forests. Breeding birds characteristic of this community type include the barred owl (*Strix varia*), wood duck (*Aix sponsa*), red-shouldered hawk (*Buteo lineatus*), great blue heron (*Ardea herodias*), green-backed heron (*Butorides striatus*), pileated woodpecker (*Dryocopus pileatus*), Louisiana waterthrush (*Seiurus motacilla*), Acadian flycatcher (*Empidonax virescens*), northern parula (*Parula americana*), prothonotary warbler (*Protonotaria citrea*), cerulean warbler (*Dendroica cerulea*), tufted titmouse (*Parus bicolor*), and yellow-bellied sapsucker (*Sphyrapicus varius*).

THREATS: This community is subject to frequent catastrophes by flooding. Fires formerly burning from the uplands crossed wet meadows and marshes and destroyed the floodplain forest edge, permitting invasion by river birch or swamp white oak. Without further fire, this forest was invaded by bur (*Quercus macrocarpa*) and white (*Q. alba*) oak, silver maple, and American elm. Succession to mesic forest is rare, however, due to flooding. Since white settlement, extensive acreage has been drained, logged, and cleared for agriculture or development.

Wet Prairie

NATURAL COMMUNITIES: Rich fen, poor fen, sedge meadow, wet loams prairie, alder thicket, willow thicket.

DESCRIPTION: Wet prairies are permanently or seasonally wet areas of the prairie landscape typically dominated by grasses and/or sedges. They form a composite of low-lying grassland communities, including sedge meadows and calcareous fens. Grasses and/or sedges are the dominant plants. Often, shrubs are also present to varying degrees. The soils of these communities are typically saturated during portions of the growing season. Standing water is usually less than a foot deep at any time of the year. Water accumulation is a result of surface runoff or subsurface discharge (as is the case in fens). In these lowland areas, sedge clumps or "hummocks" can often be observed. Wet prairies are typically located in river valleys, around the borders of marshes and lakes, and in depressions in glacial till.

DISTRIBUTION IN IOWA: Statewide.

PLANTS: Common plant species include swamp milkweed (*Asclepias incarnata*), panicled aster

(*Aster lanceolatus*), field horsetail (*Equisetum arvense*), spotted joe-pye weed (*Eupatorium maculatum*), northern bedstraw (*Galium boreale*), American bugleweed (*Lycopus americanus*), swamp lousewort (*Pedicularis lanceolata*), Virginia mountain-mint (*Pycnanthemum virginianum*), purple meadowrue (*Thalictrum dasycarpum*), sedges (*Carex* spp.), fresh water cordgrass (*Spartina pectinata*), bluejoint reedgrass (*Calamagrostis canadensis*), fringe-top bottle gentian (*Gentiana andrewsii*), yellow stargrass (*Hypoxis hirsuta*), New England aster (*Aster novae-angliae*), and cattail gay-feather (*Liatris pycnostachya*). Fens possess a distinctive group of calcium-loving plant species due to the high levels of calcium and magnesium dissolved within the water. These include tussock sedge (*Carex stricta*), marsh muhly (*Muhlenbergia glomerata*), Carolina grass-of-parnassus (*Parnassia glauca*), shrubby cinquefoil (*Potentilla fruticosa*), lesser fringed gentian (*Gentianopsis procera*), swamp birch (*Betula pumila*), and hoary willow (*Salix candida*).

BIRDS: Animal species preferring wet or hydric conditions inhabit the wet prairie. Birds of this community include the American bittern (*Botaurus lentiginosus*), sora (*Porzana carolina*), sedge wren (*Cistothorus platensis*), and common yellowthroat (*Geothlypis trichas*).

THREATS: Historically, wet prairie communities were fire-maintained. With the suppression of the fire regime, wet prairies are encroached upon by dogwood (*Cornus* spp.), willow, and other woody plants. Habitat destruction through drainage, cultivation, and excessive grazing has also destroyed much of this community type.

Open Water

NATURAL COMMUNITIES: Lake, pond, river, stream.

DESCRIPTION: Open water habitats are communities of permanent or semipermanent open water dominated by submerged aquatic plants. They are separated from marshes by a general lack of emergent vegetation. Open water habitats are found in most portions of Iowa and include rivers, backwater sloughs connected to major rivers, lakes, and ponds.

DISTRIBUTION IN IOWA: Statewide.

PLANTS: Species composition and distribution are dependent on water depth, chemistry, temperature, clarity, water movement, and the nature of the substrate. Common plant species include American lotus (*Nuphar lutea*), tuberous water-lily (*Nymphaea tuberosa*), pond weed (*Potamogeton* spp.), common water-flaxseed (*Spirodela polyrrhiza*), duckweed (*Lemna* spp.), smartweed (*Polygonum* spp.), slender naiad (*Naias flexilis*), common hornwort (*Ceratophyllum demersum*), and water milfoil (*Myriophyllum* spp.).

BIRDS: Although no bird species nest in open water, many do utilize this habitat for feeding purposes. Species utilizing this habitat and nesting in the state include belted kingfisher (*Ceryle alcyon*), double-crested cormorant (*Phalacrocorax auritus*), and great egret (*Casmerodius albus*). During migration, diving ducks and other waterfowl abound in open waters, feeding on fingernail clams and other freshwater invertebrates. These include common goldeneye (*Bucephala clangula*), lesser scaup (*Aythya affinis*), canvasback (*A. valisi-neria*), redhead (*A. americana*), bufflehead (*Bucephala albeola*), tundra swan (*Cygnus columbianus*), common merganser (*Mergus merganser*), horned grebe (*Podiceps auritus*), American white pelican (*Pelecanus erythrorhynchos*), and common loon (*Gavia immer*).

THREATS: Natural open water habitat in Iowa has suffered dramatically since white settle-ment. Many of the larger open water marshes and kettleholes were drained for farming practices. Others have been degraded through siltation and other forms of pollution. Many of Iowa's rivers have been negatively affected by channelization which effectively destroys once-available habitat. Within the past half century, reservoir construction activities have added many acres of this habitat type to the state.

Iowa
Birdlife

Common Loon

The "maniacal laughter" call of the Common Loon is one of my fondest memories of the vacations spent fishing in northern Minnesota in years gone by.

With the filling of Lake Red Rock, I expected to see loons quite often during spring and fall migrations but such has not been the case. Seldom does a loon stop at Red Rock. I have seen only two, one in the flooded Pinchey area of the refuge in October 1973 and one in November 1970 on the tailwaters below Red Rock Dam. A few have been seen on the lake by others, but they are rare here.

Worldwide there is only one family of loons with but four species. All four are found in North America, but only the Common Loon regularly migrates through Iowa after nesting on lakes in the northern states or Canada. The Arctic Loon and the Red-throated Loon are seldom seen in Iowa. Woodward H. Brown, in his "Annotated List of the Birds of Iowa," rates them both as accidentals here. Birders are advised to check the identity of all loons carefully, however.

This one on the tailwaters was a joy to watch. Its 24-inch-long torpedo-shaped body was attired in winter plumage, brown above and white underneath, with a long, heavy, dark bill and red eyes. Occasionally, it stood on its tail at full height and flapped its wings. It did a thorough job of preening, rolling over on its side, the white breast and belly gleaming in the sunlight as it dressed the belly plumage with its bill.

Dressing the feathers with oil from a gland located above the tail is a necessity with waterbirds, keeping them buoyant.

The loon feeds largely on small fish which it pursues beneath the surface of the water with great power and speed. The smaller mergansers move quickly out of the way too. A loon can swim underwater for considerable distances, often escaping danger in this manner. During summer it takes many frogs and also eats aquatic plants.

In the nuptial plumage in spring, the loon is quite a handsome bird with a dark green head, fine black-and-white stripes down the neck, black-and-white cross-banded back, and speckled sides.

Loons need solitude on their nesting grounds. Their numbers are declining partly due to the ever-increasing traffic of noisy motorboats disturbing them and causing them to abandon their nests. Pesticides are also contributing to the decline.

They are extremely awkward on land and seldom leave the water except to nest. A large circular mass of wet soggy weeds and other plant material serves as a nest, heaped up in shallow water or on the ground close to water. The two, rarely three light olive or olive-green eggs sparingly marked with small spots of brown are incubated in about twenty-nine days. The loon chicks, covered with dark wiry fuzz, are able to dive and swim well in a day or two after hatching, catching fish and showing remarkable progress within a week.

The loon gets its name from its call, which does sound like maniacal laughter. It does, however, have at least four distinct calls: a short cooing note often heard when in a flock, a long-drawn-out note known as the night call, the laughing call, familiar to everybody who has ever been in loon country, and another call not often heard, which the guides call the storm call, heard only before a storm.

Pied-billed Grebe

Iowa's first snowstorm of the season on Monday, November 13, 1972, was a horror for motorists, farmers, livestock, and wildlife, coating unharvested soybeans and corn, as well as weed fields, with ice and clinging, wet snow.

At four the next morning, the snow had stopped falling. I shoveled a path to the feeders and scattered a gallon of cracked corn on the ground, in roofed feeders, and in the galvanized "pig-creep" self-feeder. At break of

day the birds flocked in and continued flocking in to feed all day long.

Window-watching was pure joy. Too many Starlings, of course, but with nine suet feeders the woodpeckers, nuthatches, and chickadees had their chance. The tiny Red-breasted Nuthatch not only ate its fill but spent much of the day "pack ratting" bits of suet to stash in the bark of nearby trees. An uncommon winter resident in Iowa, it had been a constant visitor to my suet feeders since November.

On Thursday morning the town marshal, Arthur Anthony, knocked on my door. Slightly apprehensive, though I could recall no recent nefarious activity on my part, I was vastly relieved to learn that Tony had brought me a small dark bird he had rescued from the snow at the base of the tall city water tower.

Sunk into the snow from the body warmth and the fall, it was a wonder it was ever found, but the snow broke its fall and saved its life after its collision with the tower. (Tall structures, especially the very tall TV towers, have increased the hazards of bird migration a thousandfold.)

Gladly I accepted Tony's gift of the storm, a Pied-billed Grebe, and a very sassy grebe it was, pecking at my hand repeatedly as I put it in a bird cage which did not please it one little bit. It flailed around awkwardly on the cage floor.

The grebes are divers with legs set so far back on the body that they are unable to walk on land and totally unable to take flight unless on water. They belong to the order Podicipediformes, meaning the "rumpfoots."

I could not bear to see the little bird so uncomfortable. The remedy was at hand, an antique copper wash boiler and a couple of buckets of water. As I transferred it I checked wings and legs—thank goodness, no broken bones. On the water it was entirely at home, swimming vigorously from one end of the boiler to the other with its peculiar lobed toes.

My next problem was what to feed the grebe. On Friday I phoned Wesley Newcomb, U.S. game management agent in Des Moines, and he suggested fish pellets used at fish hatcheries if I could not get minnows. The sensible and kindest solution, Wes said, was to release the grebe, if it had recovered, on a nearby farm pond.

That evening I made arrangements with my sister, Mrs. Ray Proffit, for Ray to pick up the bird Saturday morning for later release on their pond within sight of the farm home. My sister picked up the grebe Saturday morning and released it on their pond. She reported it went into the water in a beautiful dive, stayed under a long time, and finally surfaced on the far side of the pond. The rest of the day she enjoyed watching from the window as the grebe floated, rested, preened, dived, and fed. By evening it took off to continue its southward migration, hopefully bypassing water and TV towers.

The grebes and loons are the most primitive birds on earth. Worldwide there are only eighteen species of grebes, all belonging to one family.

The Pied-billed Grebe, 9 to 10 inches long, is a drab brownish gray diver with a plain gray chicken bill and a white throat in winter. In summer the bill is white with a black band across the middle (hence the name pied: two or more colors) and the throat is black. The wings are very short but strong.

This dabchick is found from the Canadian provinces to Argentina and is the most widely distributed of the six American grebes—the other five are western. A solitary bird never seen in big flocks, it has been recorded in Iowa every month of the year, but it is rare in winter.

It breeds in marshy areas throughout the northern portions of its range and is a common breeding bird in Iowa. All grebes build floating nests of old vegetation, but the Pied-bill makes its more substantial by adding mud from the bottom of the pond. They lay five to nine deep buff eggs. On leaving the nest during incubation, the eggs are covered with nest material, which serves a dual purpose of concealment and temperature regulation.

The cute little striped chicks find safety by riding on the parental back. To teach them to swim, the parent simply dives from under them.

Grebes dive for their food, mainly fish. Usually a mat of feathers is found in their gizzards, perhaps to serve as a filter by holding sharp fish bones until they soften and can pass through the intestinal tract without puncturing it.

Carl Kurtz

White Pelican

So rare and exciting is a visit of the White Pelicans to Red Rock that we send word up and down the valley alerting all those interested in birds to "come see." On September 19, 1977, I counted 150 pelicans east of the Mile-long Bridge. Early the next morning a Knoxville group came to see them before going to work. Later that morning a group of Pleasantville birders and I enjoyed the fishing, loafing, flying, and preening of the huge birds.

Another flock flew in that afternoon, bringing the total to well over 350. Later that week Jon Stravers of Pella counted 420.

The majestic White Pelican is the largest bird reaching Iowa, with a wingspread of 9 to 10 feet, body length of 50 inches, and weight of 15 to 20 pounds. They require a daily diet of 6 to 8 pounds of small fish, mostly gizzard shad and carp. They are white all over, except the primary wing feathers and half the secondaries which are black. The enormously long bill, pouch, and feet are yellow.

The pelican is one of the oddest and one of the oldest birds on earth, a bird of great antiquity. Our White Pelican is very similar to the one found in Europe although separated by some 8,000 miles. Ornithologists say that common ancestry implies at some time continuity of range. Around 30 million years ago a mild temperate or subtropical circumpolar climate existed, and the White Pelican inhabited the shores of the Arctic Ocean. The Ice Age forced this polar pelican southward, some into Europe and some into America according to Roger Tory Peterson.

Of the eleven pelican species found throughout the world, only two, the White and the Brown, are found on this continent. Normally, only the White reaches Iowa, but there have been two or three records of sightings of the Brown.

Their stay with us is usually short, seldom more than a week, occasionally a month. This flock of 420 is the largest we have ever enjoyed at Red Rock. The previous high was 70 in September 1970 for a period of one week. They were followed by a flock of 12 that fished the shallow waters near old Highway 14 for almost a month.

Seventeen were here as late as June 9, 1971, probably subadults as they do not breed until they are three to four years old. One subadult spent July 1972 on the lake, also one during July and August 1974. Two to six were present the summers of 1975 and 1976 but none in 1977.

Pelicans will go to great lengths—and great distances—to ensure a safe breeding place for their homely leathery babies even if it means a round-trip of 100 miles to obtain the necessary food. In Great Salt Lake for years 10,000 White Pelicans (far fewer now) nested in safety on Gunnison Island, but they paid dearly for that safety, traveling 50 miles each way to obtain carp in Bear River National Wildlife Refuge in north-central Utah.

Most of the pelicans we see here probably nest on islands in the lakes of the prairie provinces of Canada, often in company with their cousins, the Double-crested Cormorants.

H. Albert Hochbaum, renowned for his waterfowl research at the Delta Marshes of Manitoba, writes, "White Pelicans nest on islands in Lake Manitoba, making daily trips to the Delta Marsh to forage for food. When they are finished with their meal, they climb the thermals at the south edge to play in wide wheeling circles until, specks in the sky, they turn and glide home to their stinking islands." They overfeed their babies, leaving the surplus to rot, attracting swarms of flies so that a pelican island can easily be located at a distance by the black cloud of flies above it.

Blue-listed since 1972, the drastic population decline is due to persistent pesticides, reduced habitat, and illegal shooting. Nesting always on remote lakes in Canada, North Dakota, and Colorado and on the Great Salt Lake and Salton Sea, they cannot tolerate human interference on the breeding grounds. Their favorite island in the Salton Sea, used since 1907, was taken over in recent years by the Atomic Energy Commission as a testing ground. Powerful motorboats now reach once inaccessible islands, and too many sightseers also cause desertion.

Scientists doing research are very careful not to linger long at the nest sites. Dr. Norman Sloan, Michigan Technological University, has been studying the birds for many years. On an island in Chase Lake National Wildlife Refuge in North Dakota, he banded 1,200 in 1972, some with yellow leg bands. Some were dyed yellow (with a harmless dye) on wings, tail, and breast. A request was then made in ornithology journals for reports to determine migration routes and winter grounds.

One evening I walked the trail through the woods in North Elk Rock State Park to the top of a high sandstone bluff, and there I could look down on the big birds on nearby sandbars. Some were swimming abreast fishing, occasionally opening the beak wide and sticking it straight down into the water to catch a fish. A few were in flight, coming quite close to the cliff, the flap, flap of the huge wings loud and clear. By 6:30 P.M. all were loafing, standing on logs and driftwood, meticulously grooming their plumage, opening and closing the beak, nibbling at the feathers, arching the neck gracefully to preen the breast feathers.

At 6:45 P.M. they lifted gracefully in the air, formed into long straight skeins of thirty to fifty, and flew over the top of Elk Rock Bluff, then dropped down out of sight to their night roost on the waters of White Breast Bay.

Scores enjoyed the birds during their stay on the lake. Herb Dorow of Newton got some excellent pictures from the cliff, while Jon Stravers and Carl Kurtz canoed within 30 yards of the flock to get close-ups. It was a field day for photographers and birders alike.

Carl Kurtz

Magnificent Frigatebird

A frigatebird in Iowa? "Ridiculous!" I said, when notified of one at Clear Lake in September 1988.

It was true, and many Iowa birders made a mad dash to Clear Lake. Most were lucky and had a good look at this "Man of War" bird.

As usual, disliking travel, I stayed home with the facetious remark, "I'll wait for one at Red Rock." And then on Sunday night, I had a long-distance phone call from Ann Johnson of Indianola, informing me that Dawn DeVore, a Simpson College junior and part-time naturalist at the Red Rock Corps of Engineers' Visitors Center, had seen a frigatebird, a female, from the window that day, October 2, 1988.

Monday morning, October 3, with lunch, a deck chair, and a spotting scope, I drove to Red Rock Dam where I set up my equipment beside the butterfly garden on the north side of the center. I was viewing the full length of the dam, the lake, and the south bluff with trees.

While I watched, an adult Bald Eagle, perched in a big tree atop the bluff, would swoop on the flock of coots on the water below. I decided it was harassment and that it had probably eaten a fish breakfast early that morning.

I glanced toward the north end of the dam and held my breath. There came approximately one hundred Ring-billed Gulls circling around and around, riding the updraft of air caused by the wind hitting the downside of the dam.

They were about 100 feet above the dam, kettling almost like hawks. This I had often observed, but the long, thin black bird with the sharply angled wings and the scissor tail circling just above the gulls with wingspread nearly twice that of the gulls was, indeed, a frigatebird and a new lifer for me.

I watched this show as it came closer and closer and closer to the south end of the dam where I was sitting. Never once did the frigatebird flap a wing, but now and then he opened or closed the scissor tail. Finally, they flew downstream and I lost sight at 10:30 A.M.

My bird was all black, a male, and I wondered if anyone would accept my identification. Dawn's bird had a white breast, definitely a female. Later that day, David Youngblut of Indianola observed this male and from the dam its flight west to the marina, and he identified it as a male frigatebird.

I rushed to Central College in Pella to tell my longtime friend, biology professor John Bowles, that I had watched a frigatebird and ringbills in a spectacular display. He was amused at my excitement. After all, he had seen many in his years in the Orient.

Why did Iowa have three Magnificent Frigatebirds in September and October of 1988 when the last (doubtful) record was 1903? In proportion to their wingspan, they are the lightest weight of any group of birds. Hurricane Gilbert, the worst ever known in the Gulf, hit northern Mexico on September 16, 1988, carrying the birds inland to several states besides Iowa.

Tom Kent, in an in-depth article in *Iowa Bird Life*, wrote that "inland records are mostly from August through October with a peak in September. These findings suggest three different patterns of vagrancy: birds of varying age and sex wandering north along the coast during warm months; 'immatures' wandering north along the West Coast in late summer; and predominately adults being carried inland during the hurricane season."

Carl Kurtz

Great Blue Heron

About the first of August the Great Blue Heron begins its postbreeding migration northward (the reverse of most birds) to signal the start of fall migration.

The Great Blue is Iowa's tallest bird, one of the most stately of all birds, standing over 3 feet high and with a 6-foot wingspread. It is indeed a picture, standing silently in shallow water awaiting an unwary fish to spear with its long beak or flying with slow, easy strokes with its great wings. It is easily distinguished from other birds by the down curve of the wings. The Bald Eagle's wings in flight are straight across, and the Turkey Vulture's are uptilted in a shallow V.

For nine years since the filling of Lake Red

Rock in 1969, I have enjoyed these birds, especially during the fall when the population sometimes exceeds 250. And each year we have a few of the Great Egrets, which have white feathers, and the southern Little Blue Herons, which are also white until the second year. Rarely, we have a Cattle Egret roosting with the other herons after it has spent the day chasing grasshoppers in nearby cattle pastures.

I recall that fall of 1969 when John Beamer, then Red Rock management biologist, and I discussed the possibility of a rookery as we watched at least one hundred of the Great Blues feeding around the potholes on the floodplain below refuge headquarters. I thought possibly 1971 might be the year for such a rookery, but John was sure we'd have them nesting there the next year and he was right.

So in April 1970, twelve pairs of the big birds went through their ritual courtship dances, selected mates, and began gathering big sticks and building their huge nests in the tops of tall trees standing in shallow water on the floodplain just west of Red Rock Bluff. Good-sized branches were broken off and carried to the nest site, and always one bird, probably the female, remained standing at the nest to receive the stick, then carefully worked it into a nest. The completed nests were flat structures 30 to 40 inches across.

Four to six green eggs, larger than duck eggs, were laid in each nest, and then the normal twenty-eight-day incubation period began. Both the male and the female took turns incubating, and it was delightful to watch the ritual of changing places on the nest, then the maneuvering necessary to get their long legs folded and down on the eggs.

Unhappily, I discovered on May 22 that three of the nests were deserted. As I watched with a 20-power scope, I could see that eggs were breaking, and my heart sank. The persistent pesticides upset the calcium metabolism, causing the females to lay thin-shelled eggs. By May 23 all nests were deserted.

I reported this in July of 1970, mentioning the deadly effects of DDT and dieldrin. At that time the literature was full of reports on the sublethal and lethal effects on wildlife. Britain already had banned dieldrin.

In 1971 the rookery had grown to thirty-five nests, all failures—not a single heron chick and still no research.

In late June 1972, a *Des Moines Register* writer, Patrick Lackey, and photographer, George Ceolla, asked me when we'd have heron chicks in the forty nests. I replied that there would be none, that most of the nests were already deserted, and that I had failed to get anyone to research it. I finally advised them to telephone Dr. David Trauger at Iowa State University and suggested perhaps he could get a research biologist interested.

Finally, Dr. Larry Wing of ISU, with the aid of a helicopter crew, collected a few of the last eggs. The chemical analyses by John Richard revealed a horrifying load of persistent pesticide residues in the embryos.

The next year, 1973, was a big flood year, and fortunately for the herons, almost all the nests were destroyed. The herons then established their rookery at Lake Rathbun, which has a much smaller watershed with fewer farm chemicals polluting it, and there they have successfully reared young herons each year.

Linda Kurtz

Great Egret

During the last weeks of August and the first weeks of September 1982, I enjoyed a ballet performance for an hour before sunset and half an hour after. The ballerinas were the Great Egrets, at least one hundred of them—snowy white, the most ethereal of all the egrets, standing almost 3 feet tall. I had a panoramic view from the cliff-top Conservation Commission parking lot south of Runnells.

On the floodplain many pools of shallow water were interspersed with mud bars. The willows formed a lacy backdrop, and in the distance were the dark green, wooded bluffs. It was on one of the mud bars that the Long Whites congregated to dance and preen and later to roost.

Some evenings only a few danced and on other evenings, many. The dance is really a 4-foot jump into the air with wings flapping and long black legs dangling, then down again touching ground lightly in almost the same spot.

We are lucky to be able to see these birds today. In the early 1900s they were near extinction. Called the Long Whites by the Florida plume hunters, they had been hunted since 1840 for the long graceful plumes that extend from the head down the back and beyond the tail like a bridal train.

These nuptial plumes, called aigrettes, appear only during the mating season. The plume hunters killed the birds at their colony nests, took the plumes, and left the three to four baby egrets to starve to death.

The plumes of four birds weighed 1 ounce, sold for $32, and were used in the millinery trade to decorate women's hats. This made their worth about twice the value of gold at that time. In 1902, the London Commercial Salesroom had 48,240 ounces of plumes, which meant that 192,960 egrets had been slaughtered at their nests.

A vigorous educational and legislative campaign by the National Audubon Society and many other organizations and individuals finally resulted in complete protection of all the herons and egrets, but not before one Audubon warden had been shot and killed in Florida.

As a child and young adult I went on many visits to my grandfather in his cabin on the Des Moines River near Bennington Bridge but never saw a Great Egret. It was not until World War II that I made my first sighting in the valley.

Slowly their population increased, only to suffer another decline caused by DDT and dieldrin. Since the ban on these pesticides went into effect, the egrets have recovered somewhat, and in the fall of 1981 more than 120 were counted east of Mile-long Bridge.

There probably were more this fall of 1982, as they were scattered all through the 26,500-acre Red Rock Refuge, fishing in shallow waters. However, many winter in Central and South America and the Caribbean Islands, where there are no bans on pesticides.

Some egrets nest in colonies along the Mississippi River and in some northern states, but there are no nesting colonies in interior Iowa. Most nest in the southern states and are noted for their postbreeding wandering in many directions.

Audubon in 1840 described their spectacular courtship. "The males, in strutting around the females, swelled their throats, emitted gurgling sounds, raising their long plumes almost erect, and paced majestically before the fair ones of their choice.

"Although these snowy beaux were a good deal irritated by jealousy, conflicts now and then took place, but with less fighting than among the Great Blue Herons. These meetings took place about ten o'clock in the morning on a sandbar and continued until nearly three in the afternoon, then all flew off in search of food. These maneuvers continued nearly a week."

The big stick-nest from 15 to 80 feet up in a tree is built by both sexes. Usually the male brings the sticks and the female places them in the structure. The pale greenish blue eggs, usually three to five, are incubated twenty-five to twenty-six days by both sexes, and the nestlings are fed a fish diet by both parents. At three weeks, they are "branchers," walking around on branches near the nest, and at five

weeks fly short distances. At six weeks they fly with the parents.

Carl Kurtz

Cattle Egret

Naturalist Larry Totten is a miracle man. On a Des Moines Audubon Society field trip in September of 1977 at Jester Park just north of Des Moines he really delivered. At the beach along Saylorville Reservoir we were surprised and delighted to see four rare Cattle Egrets on the nearest weedy sandbar busy gleaning grasshoppers. All four were young birds with yellow bills and dark legs. Adults have orange bills and yellowish to pinkish legs. The bird is 17 inches long with a 3-foot wingspread. In spring breeding plumage, the crest, breast, and shoulders are buffy-orange. In fall they are white all over.

Binoculars were adequate for excellent viewing, but a couple of spotting scopes were set up for more detailed viewing. The egrets didn't need to go to cattle pastures because the filling of the reservoir had concentrated the grasshoppers on the few sandbars.

At Red Rock Reservoir we waited more than a year before sighting our first Cattle

Egret on October 19, 1970. That one, also immature, was feeding on grasshoppers with the cattle on the James Rees farm at the south edge of Red Rock Refuge.

I rushed back to refuge headquarters and brought John Beamer, our management biologist, to confirm my identification. (Rare birds should be identified by at least two experienced birders or by a photograph which is proof positive. Also, a rare bird documentation form must be filled out in detail and mailed to Dr. Nicholas Halmi who is editor of the Iowa Ornithologists' Union Field Reports.)

These four egrets brought back fond memories of the excitement in the ornithological world during the early 1950s when the Cattle Egret was first discovered along the Atlantic coast. At that time we lived in Georgia so had an early view of the bird during trips to Hilton Head Island, South Carolina, and Savannah Beach, Georgia.

The Cattle Egret was the very first Old World bird to reach the United States in recent times on its own power and to become established as a breeding bird. It came from Africa by way of South America in 1930 and reached Florida in 1952. The 1,770-mile trip from Africa to South America sounds impossible, but they could have rested on St. John's Rocks in the equatorial Atlantic Ocean, as one was seen there in 1963.

The Cattle Egret was quickly given United States citizenship by placing it on the list of birds protected by the Migratory Bird Treaty Act and giving it AOU (American Ornithologists' Union) Number 200.1. Thereafter it was illegal to import, cage, sell, kill, collect, or molest the birds, their eggs, or their nests.

The first Cattle Egret in Iowa was observed and collected by Dr. Milton Weller (then professor at Iowa State University) April 21, 1961, incidental to fieldwork on marsh ecology. Currently Dr. James Dinsmore, professor of animal ecology at Iowa State University, has begun compiling a survey of the Cattle Egrets in Iowa, past and present, and thus one should always record the number of birds, date, and location. If possible have a birder friend confirm the sighting or take a picture.

All other herons feed on fish and amphibians, taking insects only occasionally, but the Cattle Egret is plainly insectivorous, walking alongside a cow and spearing the insects stirred up by the grazing animal.

The Cattle Egret's feeding habit is a classic example of commensalism, that relationship in which animals live in close and harmless association with one another, from which only one partner derives benefit. I think the cattle and the farmer also derive benefit. We may need to redefine commensalism, as I have several times seen the Cattle Egret following a tractor in spring instead of a cow. A number of factors may account for the population explosion of the Cattle Egret since 1952. A colonial nester, it moves into rookeries with our native herons, but it builds a stronger nest 4 to 6 feet up in small shrubby trees in or near water. It nests in May in the United States and in October in South America.

Also it breeds when one year old; most other herons breed at two years. The full clutch of blue eggs numbers four or five, with incubation beginning with the first egg.

The nestlings are well fed on insects, even though the adults at some rookeries in Georgia must travel 9 to 13 miles to cattle ranches to obtain food. They now nest in every coastal state except in the Northeast and Northwest; 20,000 pairs were in Texas by 1965, in California, and along both coasts of Mexico. They have moved north into Tennessee, Arkansas, and two sites in Missouri, one being near St. Louis. On the inland coastal plains area, there were 1,500 nests in the Rebecca, Georgia, colony. And most surprisingly, they now nest in two locations on islands in Ontario, Canada.

In a recent telephone call, former Des Moines Audubon Society member Bill Criswell, now in Dyersburg, Tennessee, reported counting 1,134 Cattle Egrets in the heron roost on the south edge of Dyersburg. There is no food competition with our native herons since the insectivorous Cattle Egrets have exploited an essentially vacant ecological niche, one in part created by the cattle industry.

A. Cruickshank/Vireo

White-faced Ibis

If by chance—and it will be a remote chance—you come upon a most peculiar bird, one with a 5- to 6-inch-long, thin, down-curving "droop-snoot" beak, do not be alarmed. It is not sick, nor is it deformed. It is a White-faced Ibis, and it is well adapted for its special feeding habits.

White-faced Ibis sightings are rare, about eight or ten every ten years. Statewide, only seven were seen between 1954 and 1964—two in Johnson County, one each in Monroe, Hamilton, Emmet, and Greene counties, and one on Credit Island. Several pair nested near Spirit Lake in 1986.

After Red Rock filled in 1969, we had no ibis sightings until September 1981, when Jim Sinclair, president of Rolling Hills Audubon Society of Indianola, spotted one at the west end of Red Rock Refuge near Runnells.

Then on May 25 and 26, 1982, Corps of Engineers ranger Jerry Dowell observed one at close range in a marshy area north of the tailwaters below Red Rock Dam. The bird was not at all alarmed when Jerry slammed on the brakes, came to a halt, and then sat in the car noting all identification marks for the necessary documentation of a rare bird.

The ibis is about the size of a small heron but with shorter legs. It is about 19 inches long with a 3-foot wingspread. The body is a beautiful glossy deep reddish bronze, which looks black at a distance. The long bill is dark. There is a white line of feathers around the eyes and under the chin of the adult.

Unlike the heron, the ibis flies with bill, neck, and legs outstretched, with rapid wingbeats of the short, broad wings, flapping and gliding alternately. Flocks fly in diagonal lines or in bunches. Sometimes they mount high in the air in broad circles, then plunge downward with legs dangling. The call, seldom heard during migration, is a nasal "ooh-ich" or a hoarse "ka-onk, ka-onk."

The diet consists of earthworms, crayfish, small mollusks, insects and their larvae, small fish, frogs, and possibly some aquatic plants.

They feed along riverbanks and in shallow edges of pools, ponds, and marshes.

Bent, in *Life Histories*, reports that during the late nineteenth and early twentieth centuries huge colonies nested in tules in the "hog-wallow prairies" of Texas along with egrets and night herons. Those swamps were drained long ago. Large breeding colonies of two hundred pairs were reported in San Jacinto Lake in Riverside, California, in 1911.

Dr. Thomas Roberts, in his monumental two-volume *Birds of Minnesota*, records, "On June 16, 1894, two nests and six adults seen, and on June 22 and July 2, 1895, two sets of 3 and 4 eggs at Heron Lake in southern Minnesota in the midst of a Black-crowned Night Heron colony.

"The nests were among reeds and rushes compactly built of pieces of reed stems, rushes, and grasses, supported above water, more cupped than heron nests, eggs deep bluish green, incubation 22 days. On fledging, the young were glossy green above, dark below, little or no chestnut and lacked the white face."

In the March 1982 issue of the National Audubon Society's *American Birds*, I found this encouraging note. In the fabulous Malheur Wildlife Refuge in Oregon, "White-faced Ibis continued its phenomenal nesting success with 650 pairs, up from 190 in 1978."

Some taxonomists believe that the Eastern Glossy and the Western White-faced Ibis are conspecific and should be listed as a single species, the Glossy Ibis.

The Glossy is found only along the coast from Florida north to New England, and a recent note in *American Birds* reported a drastic population decline of Glossy breeding colonies in South Carolina over the past four or five years.

Prior to 1915, there was an open season on this species with a bag limit of twenty in California. They were called Bronze Ibis or Black Curlew because of the down-curving bill similar to the curlew's.

Larry Stone

Giant Canada Goose

Each summer a few fortunate Iowa wildlife biologists go to Churchill on Hudson Bay to study Canada Geese on the breeding grounds. I envied them in past years but not during the summer of 1975, for I had the pleasure of observing a pair of Giant Canada Geese rearing their five goslings at a nearby game farm.

Three races of Canada Geese reach Iowa during migration. The smallest and rather rare is *Branta canadensis minima*, not much larger than a Mallard duck. It nests much farther north in the Arctic than *B. c. interior*, our common Canada seen here in large flocks. *Interiors* are big, but a 10-pounder is a whopper.

The Giant Canada Geese, *B. c. maxima*, weighing up to 18 and 20 pounds, were once the breeding geese of the prairie biome from the prairie provinces of southern Canada south as far as Reelfoot Lake in Tennessee, where they often nested in trees, east to Michigan and west to the Continental Divide. For thirty years the Giant race was thought to be extinct in the wild until Dr. Harold Hanson, Illinois Canada Goose authority, at the invitation of Minnesota biologists, investigated the Canada flock wintering on the open water below the Rochester, Minnesota, power plant.

On a subzero day in January 1962, Hanson and the local biologists began weighing, measuring, and banding. The weights were so great they decided the scales were faulty, so they sent one man to purchase a 5-pound bag of sugar and a 10-pound bag of flour.

Hanson said, "A quick test of the scales revealed that the 'impossible weights' we had been getting were correct. Now we knew beyond question that we were dealing with a very large race." In fact, they had found the Giant race so long considered extinct.

My observation of the Giants began when Jim Davis, owner of the private game farm and half-mile-long lake, and Bill Burrell, owner of a lake home there, gave me permission to observe the dignified protective Giant Canada parents, their five goslings, and the flock of eleven nonbreeders.

They mate for life, and the courtship challenges of the gander began in March. The first nest of coarse grasses lined with down from the female's breast was broken up by marauding dogs. (Late blizzards also are a hazard—the April 9, 1973, blizzard destroyed the nesting that year.)

The second clutch of five eggs was incubated twenty-eight days by the goose, with the gander standing guard. The five goslings hatched in late May were covered with golden down except for a darker olive on the back, wings, and top of the head; bills, feet, and eyes are dark. (*Interior* goslings are much darker.)

Growth was rapid and by June 28 the goslings were handsome adolescents, three-fourths grown with typical Canada markings—snowy white undertail coverts and breast, belly and back dark brownish gray. Only the head and neck were in muted colors not fully feathered, the "stocking" dark gray woolly down instead of black, and the cheek patch a shadowy paler gray instead of white. Wing-flapping exercises revealed wing primaries about 2 inches long.

On this day the family and the always slightly distant flock of eleven were grazing in the hay fields along the north shore a quarter-mile away. At sunset all seven members of the family, with the gander leading, marched slowly and with great dignity down to the water, then walked along the muddy edge cropping arrowhead leaves.

The gander stood sentinel. Occasionally with head lowered, neck outstretched, beak open and hissing, he charged those of the flock who approached too closely. Entering the water, the family swam the length of the lake, a fifteen-minute trip to the roosting ground, a small bluegrass peninsula below the Burrell home. On landing the goslings preened vigorously, flapped their wings, and cropped a little bluegrass before sitting down, while the gander remained standing on guard.

Only once I saw the parent gander attack another gander, viciously hitting him with the spurred wing, rolling the intruder over and over. Three times the intruder tried to run to the water only to be knocked over again. Finally, limping, he made it into the water.

One evening Zoda Burrell told me the story of her empty garden on the peninsula. Earlier she had planted rows of vegetables, including sweet and hot peppers, cabbage, tomatoes, and onions. One morning she looked down at a bare garden nearly cleaned out by the geese—everything gone except the tomatoes and onions.

She did not replant the vegetables; thereafter it was goose-roosting territory. The geese were more fun to watch. However, Bill called a halt when the goslings began stripping the bark from a young mountain ash tree. He wrapped the trunk securely with layers of aluminum foil.

On July 20 I watched the family swim across the lake and climb the dry, steep south bank covered with dead brome, orchard, and bluegrass and no greenery. I wondered why as I trained the scope on them. All seven began stripping bluegrass seed heads. Turning the head slightly sideways, the stalk was grasped in the wide bill and with an upward movement neatly stripped of seeds. Hanson calls them "seed strippers" because of the wide, serrated bill and noted that in Canada the very young goslings are taught to strip seeds.

This was a first for me, although I have seen thousands of Canadas feeding in wheat and cornfields. When they returned to the water, one gosling yanked up an arrowhead plant and ate the "duck potato" root. On July 22 I observed the family with yet another food source, fishing a grassy, aquatic weed from the water near the Burrell dock. And the next morning tragedy struck. One gosling got a snagged fishhook in its throat and drowned while the anguished parents and other four goslings honked their distress.

On that same day one gosling made its solo flight halfway across the lake amid much honking applause by the proud parents and goslings. That in turn triggered a "goose roundup" by Jim and the clipping of one wing of each gosling. The migratory instinct is never bred out despite generations of game farm living.

Opportunities to observe Giants in the wild are rare, but during the winter of 1971–72 Bill Criswell and I watched a flock of nine Giant Canadas wintering on the open tailwaters below Red Rock Dam. Two, neck-banded EK and AJ, had been pen-reared the previous summer by University of Minnesota graduate student Michael Zicus at Crex Meadow Refuge in Wisconsin. On their return that spring, Zicus wrote me that EK, a gander, mated with a yearling goose and the two defended a large territory at the end of Phantom Lake all summer. Mated as yearlings, they do not breed until the second year.

Once, on October 5, 1969, I had a chance to compare the extremes in sizes of the Canadas. Over the cliff from the north a flock of ten Giants and four *minimas*, no bigger than Mallards, flew in and landed on a Red Rock Refuge wheat field.

Larry Stone

Wood Duck

Those Beau Brummells of the duck world, the Wood Ducks, arrived in the refuge in early March. Attired in glorious nuptial plumage, they very quietly went about their usual spring routine, at first just feeding, loafing, and preening, later househunting and courting.

As I toured the south side of the Red Rock Refuge there were Wood Ducks everywhere, not only within the refuge but on every nearby farm pond. And they multiplied mightily, for by late summer, John Beamer, refuge manager, counted three hundred along the river, and he believes at least half were reared here, as little groups of ten to twenty ducklings were seen at numerous points along the river during the breeding season.

The plumage of the drake is the most brilliant of any North American waterfowl. The crest of his head is iridescent green, blue, and purple, with two white stripes. The eyelid is bright red and the iris red or brown. The bill is pink, white, yellow, and black. The throat is white with two white points up the side of the head and neck. The maroon breast is marked with small white triangles. The dark back and tail are iridescent green and blue.

The buffy sides boldly bordered with black and white are streaked with fine black lines. The belly is white and the feet dull yellow.

The drake's eclipse plumage, lasting from June through August, is subdued much like the female's, but the bill and eyes retain their bright colors. The big wing feathers are shed in July or August and he is flightless then, spending his time on the water in remote wooded creeks or rivers.

The female is light gray-brown, with iridescent green and purple on back and wings. The crested head is gray; throat and feathers about the eye are white. The bill and feet are gray or olive. It is the only female duck having iridescent plumage on the body.

The voice of the drake is rather unducklike—a whistled "hooweet," often high-pitched. The female's alarm note is a short "creek-creek." The talk in the night roost is soft, low, and finchlike.

The importance of eyesight to a bird is beyond estimate. The eyes of ducks appear small but actually occupy a very large part of the head. The two eyes weigh nearly as much as the brain.

With eyes in the sides of the head, they enjoy horizon-wide vision and can take in twice as much of the landscape as humans, but people and the owl enjoy the advantage of binocular vision with depth perception. (There is some overlapping of the duck's visual field, a narrow binocular span utilized in close-up selection of food.)

It is therefore surprising that the Wood Duck is so skillful at darting, dodging, and twisting at high speed through a maze of tangled treetops. The Wood Duck's top speed is 30 to 50 miles per hour. The Mallard or any other duck would probably knock its brains out if it tried a similar high-speed trip through the treetops. While observers hold their breath and predict disaster, the female Wood Duck shoots like a bullet with hardly a pause straight into a 4-inch nest hole. The slightest miscalculation would be fatal—pressed duck!

The Wood Duck is Iowa's only hole-nesting duck (except for the rare Hooded Merganser). They prefer trees near water but often nest in trees far from water in farmyards and in town. A pair nested in Knoxville in 1970.

The height of the nest hole varies from 5 to 50 feet. The female lines the nest with down from her breast and lays one nearly round, creamy white egg each night. Full clutch is eight to fifteen eggs, and the incubation period is long, thirty to thirty-two days.

At hatching time the little Woodie, using its sharp claws, makes a complete rotation in the shell as it cuts around it with its bill. Paul Stewart reports, "Of two Wood Ducks hatching in my hands on May 7, 1967, one used 8 and the other 9.5 minutes in cutting and getting out of the shell."

Although the literature abounds with records of the female carrying the young to the ground, it is generally conceded that in response to the soft call of their mother, the ducklings use their sharp-pointed hooked claws and the hooked nail on the end of the bill to climb the wall of the nest hole or nest box, then jump to the ground, fluttering down unharmed. Observers once watched a brood of ducklings jump 22 feet and land with a bounce on a concrete sidewalk. They walked off uninjured.

The downy ducklings are a beautiful brown on the upper parts; the sides of the head, breast, and belly are light yellow. The tip of the bill is yellow, and a dark line runs through the yellow back of the eye. There is a small white spot on each wing and another on each side near the tail.

Of all the ducks, the Woodies have been most persecuted by humans. They were easily hunted because they were unafraid. Their flesh is delicious, and their feathers make good trout flies. Add to this the loss of habitat, forests cleared, potholes and swamps drained, and long years of market gunning and the Woodies soon faced the fate of the Passenger Pigeon.

The market gunners knew the ways of waterfowl and used this knowledge to decimate whole flocks. Woodies are creatures of habit and this the old gunners understood. By 1918 the Woodies were so near extinction that a closed season was declared in Canada and the United States. This protection continued until 1941 when a few states permitted a bag limit of one Wood Duck. Even today our Iowa Conservation Commission protects them as "ninety point" ducks.

So it is a pleasure to report that Woodies appear to be on the increase in several areas of Iowa. Locally, the well-managed refuge—plenty of nest holes and food—is a factor. Perhaps the most important factor of all is the recent change of diet from acorns and forest food to corn first reported by A. S. Hawkins of Illinois in 1955.

Certainly I've seen more Wood Ducks here this year than in all my fifteen years of observations in the low country of South Carolina and Georgia, the vast complex of lakes in North Carolina, Tennessee, and Alabama, and a weekly trip into the cypress, gum, and live oak swamps of Georgia during fall, winter, and spring. And the Woodies will probably stay with us until mid-November. We had two on the refuge for the 1970 Christmas Bird Count.

Carl Kurtz

Northern Shoveler

Undoubtedly the Anatidae—ducks, geese, and swans (the web-footed swimmers)—have greater universal appeal than any other bird family. Poet, prince, and peasant through the ages have thrilled to the spectacular migration of waterfowl. And two of today's princes, Philip of Great Britain and Bernhardt of the Netherlands, are deeply involved with the solution of world wildlife problems.

Peter Scott in his "Key to the Wildfowl of the World" lists 247 races within 151 full species. (Four species have become extinct within recent years.) In color they range throughout the spectrum, even pink. In voice they range from bugling, honking, whistling, barking, quacking, clucking, rattling, and grunting to laughing. In weight they range from the 20- to 30-pound swan to the little ½- to 1½-pound Bufflehead duck. There's a species fitted into every niche of the watery worlds—marshes, torrential mountain streams, rivers, lakes, ponds, estuaries, and oceans from the Arctic to the tropics.

Here in the center of the greatest of all flyways, the Mississippi, more waterfowl and all other species pause to feed and rest than at any other place on earth.

Twenty-eight waterfowl species (four very rare) reach Iowa each spring and fall. Some are divers, some dabblers, depending on their feeding methods. All ducks have rather wide, flat bills but none equals that of the Northern Shoveler. In length its bill is about 2¾ inches longer than the head, and it widens from ⅝ inch at the base to 1¼ inches near the tip.

Children always giggle at this "funny" duck until they watch it feeding, for it is indeed the king of the surface feeders, paddling with head half submerged, taking in small particles, and sifting out all unwanted material through the comblike "teeth" along the edges of the upper and lower mandibles.

Often it slurps mud from the shallow bottoms. Altogether it takes in more animal food than other dabblers, from 34 to 41 percent of its diet, including mollusks, bugs, beetles, caddis and dragonfly larvae and nymphs, other insects, tiny fish, and crustaceans. The total plant food taken—about 66 percent—includes seeds of various pond weeds, grasses, algae, duckweeds, and smartweeds. Because of the odd shovel bill, this duck has acquired more colloquial names than any other. Spoonbill or Spoonbill Duck shortened to Spoony was the name most used by early wildfowlers. A few of the local names were Broadbill, Cow-frog, Shovel-bill, Shovelnose, Souplips, Mule Duck, Scooper, and Mud Lark, my favorite.

Widely distributed, it breeds in Europe, Asia, and North America into Alaska and Canada south of the Arctic Circle. A strong little flier, it winters as far south as East Africa, the Persian Gulf, Ceylon, Burma, southern China, Japan, Hawaii, Honduras, into northern South America, and throughout the southern United States. A warm-weather duck, it is a late spring migrant in Iowa (about mid-March through April) and an early fall migrant (mid-September through October).

They seldom nest in Iowa. Woodward H. Brown, in "An Annotated List of the Birds of Iowa," lists only a few nest records in the northwest corner of the state. The nest of grass usually is hidden under a clump of grass or brush and lined with down from the female's breast. The eight to twelve olive to greenish gray eggs are incubated twenty-two to twenty-four days by the duck, with the drake deserting at the onset of incubation. The ducklings are covered with dark olive down on the head and back with pale yellow underparts.

The adult drake is handsome with a green head and neck, white breast sides, and underparts of cinnamon red-brown, back grayish brown, shoulders of the wing gray-blue separated from the green speculum by a narrow white band. The eyes are yellow, feet orange, and the spatula bill black. From July to December he wears eclipse plumage similar to the female's light tan mottled with brown. There is a flightless period in August when the primary wing feathers are molted.

Every spring and fall we enjoy them on farm ponds and in Red Rock Refuge, but this summer we had a rare treat. A drake spent all summer on the lake of the Davis game farm. The children delighted in watching old "Funny Face" feeding in the muddy water along the north shore.

Carl Kurtz

Gadwall

One November afternoon, two Pleasantville sportsmen, cousin Harvey Rees and George Chapman, brought their bag of ducks, asking me to identify one that had them puzzled. It was a Gadwall drake, a dark gray, medium-sized duck, not very distinctive when viewed on the water.

The chief identifying marks are the black upper and undertail coverts and a black posterior. The white speculum and the chestnut and black patches are also distinct if viewed through high-powered binoculars. The feet are yellow with no lobes on the hind toes, a sure mark of a surface-feeding or puddle duck. (All bay or diving ducks have lobed hind toes.) Viewed closely, the finely reticulated pattern of light gray lines on dark gray of the breast plumage was very beautiful.

It feeds by tipping, or standing on its head in the water. Much of the food is vegetable, tender grasses and aquatic plants, and occasionally it visits grainfields for wheat and corn. Because it eats very little fish and animal matter, the meat is very good. I know—I ate Harvey's Gadwall!

It is now rated an uncommon migrant in Iowa, rarely ever nesting here; the last breeding record in 1965 was at Goose Lake in

Greene County. In 1907 Anderson rated it a common migrant with nesting in Kossuth County.

The entire continental duck population is at an all-time low of 20.5 million, even lower than the bad 1930s when it dropped to 27 million in 1935. Ding Darling of Des Moines, appointed to the Department of the Interior during the Roosevelt administration, obtained cooperation of federal and private agencies, so by 1944 the duck population was 125 million. With loss of habitat and increased hunting pressure, the ducks have steadily declined for the past thirty years. Burlington's noted Wood Duck authority, Frederick Leopold, recommended five years ago that the length of the duck-hunting season be cut in half. The point system is overly generous, allowing 10 ducks a day of several species to be taken every day of the fifty-day season.

The Gadwall is an interesting worldwide species breeding in the temperate regions of North America, Europe, and Asia, from Iceland, the British Isles, Denmark, Sweden, and Holland to Kamchatka; it also breeds in southern Spain and northern Algeria. Its winter range in North America is in the southern states and Mexico; in the eastern hemisphere, it includes the British Isles, the Mediterranean basin, northern Africa to the Sudan and Abyssinia, northern India, China, and Japan.

Carl Kurtz

Canvasback

Are the hunters of America concerned about the drastic population decline of the ducks? Or are they even aware of it? Let me quote a few statistics. In pioneer days we had a population estimated at 500 million ducks. By 1900 it was 150 million. I recall our deep concern when we dropped to 27 million in 1935, and we owe a great debt of gratitude to Ding Darling of the *Des Moines Register* for reversing this downward trend through cooperation of both the federal government and private groups such as Ducks Unlimited.

By 1944 we had 125 million ducks, but the sale of federal duck stamps to hunters jumped from 600,000 in 1937 to 2 million in 1947. And so begins the really alarming decline. With increased hunting pressure, loss of habitat, and occasional drought, the duck population dropped to 54 million in 1947 and the downward trend has continued each year to 36 million by 1969 and 20.5 million in 1975, an alarming decline.

Yet the slaughter continues unabated. The U.S. Fish and Wildlife Service of the Department of the Interior is largely to blame for allowing this to continue. The states, too, are not without guilt. The scramble for license dollars and political pressure are sad facts of life. The National Wildlife Federation in February 1975 protested to the Department of the Interior its lack of action concerning the Canvasback duck population decline. Are these hunters and the Izaak Walton League members more deeply concerned than the trained wildlife biologists?

Frederick Leopold of Burlington, Wood Duck authority and brother of the late Aldo Leopold, father of modern wildlife management, recommended some five years ago that the length of the duck-hunting season be cut in half. Over the past several years, the infamous point system has contributed to the decline by allowing as many as ten ducks of some species to be shot each day, every day of the entire duck hunting season of fifty days.

The Canvasback was this year placed on the National Audubon Society's Blue List, indicating a drastic population decline. This was protested by the U.S. Fish and Wildlife Service, and sure enough, they declared an open hunting season on Cans this fall.

When Wood Ducks were threatened with extinction, they were protected by a closed season: no hunting in both Canada and the United States from 1918 to 1941. Only drastic action will save the Canvasback.

The 1976 Iowa duck stamp was a pair of Canvasbacks in flight from a painting by Nick Klepinger of Reasnor, Iowa. The money from stamps goes for research. In addition, the stamps are a collector's item.

The handsome 3-pound Canvasback has from colonial days been the favorite of hunters for its fine-flavored meat. The market gunners took a heavy toll, and this was not stopped until the Migrating Bird Treaty Act went into effect in 1918. A single market gunner often killed 150 Cans a day and 7,000 a season on the upper end of Chesapeake Bay.

Cans make up less than 1 percent of the total duck population, yet on the delta marshes in Canada, according to that renowned biologist, H. Albert Hochbaum, they made up 21 percent of the total kill in the 1938–1944 period and 27 percent of the total kill in the 1946–1950 period, determined by bag checks. The Cans have been in trouble for many years.

Dr. David L. Trauger, dedicated research biologist with the Northern Prairie Wildlife Research Center, U.S. Fish and Wildlife Service, Jamestown, North Dakota (formerly with Iowa State University), has over the past several years done research on the problems of the Canvasback, which he calls the "prestigious patriarch of waterfowl of North America." Several major factors have contributed to its decline: loss of habitat; loss of major aquatic plants on the winter grounds due to siltation and pollution of rivers and estuaries; reliance on only one food source, the fingernail clam at Keokuk and other small clams in Chesapeake Bay; ingestion of lead shot; chemical pollution; the parasitic Redhead Duck laying her eggs in Can nests; predation by raccoons; and disease, e.g., 30 percent of winter Cans on Chesapeake Bay are infected with bird malaria.

But perhaps the most alarming is overkill by increased hunting pressure and the terrific sex imbalance—only 20 to 30 percent females and 70 to 80 percent males, although at hatching the sex ratio is 51 males to 49 females. Early opening of hunting on the breeding and molting grounds in Canada has contributed to the kill of the females, which are readily attracted to small female-type decoys.

To quote Hochbaum in "To Ride the Wind," "As gunning carried on through the

1960's, a large part of the Delta Marsh seed stock—the sacred aggregation of experienced breeders—was harvested. Fewer and fewer stopped by the Delta because they no longer lived to go south. The drought ended in 1968; water, food and cover were restored in lush profusion. But there were not enough breeders to return to prime habitat. We had forgotten the fundamental rule of waterfowl management: It takes ducks to make more ducks."

He recommends a "new Migratory Bird Treaty of international controls on the early season killing of young ducks and their mothers which is equally as dangerous as spring shooting of waterfowl. Opening dates should be delayed right on down the line from Canada through Minnesota, Illinois, Iowa, etc."

To see Canvasbacks, go to the Keokuk pool on the Mississippi River in the fall: 106,000 of a total continental population of 207,000 were seen there one fall. Dredging of this feeding pool by the Army Corp of Engineers has thus far been prevented. However, one big oil spill could mean disaster since it could wipe out nearly half of the continental Canvasback population.

R. Mellon/Vireo

Oldsquaw

Notoriously noisy at all seasons, the Oldsquaw is also known as the Cockawee, Noisy Duck, and Hound. And it is not the female but the drake that is the loudmouth. The call note is a loud "ong-ong-onk."

Most of them winter along the Atlantic and Pacific coasts, but about 20,000 winter on the Great Lakes, the most numerous sea duck found there. According to Bellrose's *Ducks, Geese and Swans of North America*, a small number are seen each fall on the Keokuk pool of the Mississippi River near Nauvoo, Illinois, but they are rarities in inland Iowa.

At Lake Red Rock, one was observed on the tailwaters below the dam on January 2, 1970, and none again until November and early December 1981, when three were seen on the tailwaters.

Our Red Rock Christmas Bird Count was held last December 19, and, as usual, I was hoping we would get a rarity. So it was thrilling to hear during lunch in the Corps building that we had one female Oldsquaw on the tailwaters. We had our rarity.

Fifty other species were observed and

counted that day—320 Canada Geese, nearly 10,000 Mallards, five Pintails, 69 scaups, 28 Common Goldeneyes, two Ruddy Ducks, two Hooded Mergansers, 41 Red-breasted Mergansers, 841 Common Mergansers, 235 Herring Gulls, 1,417 Ring-billed Gulls, and so on. All very exciting, but the Oldsquaw was the thriller.

The Oldsquaw drake is a handsome bird about 20 inches long, mostly black and white with long central tail feathers, but it has a short neck, unlike the Pintail, the only other duck with long tail feathers. The female, about 16 inches long, has a white belly, brown collar, and crown and cheek marks on a white face.

As divers, they are unsurpassed. I did not carry a stopwatch but am sure the ones I watched in 1981 stayed underwater thirty to forty-five seconds—much, much longer than a nearby Goldeneye that stayed under no more than fifteen seconds.

Bellrose believes they probably dive deeper than any other duck. There are many records of them captured in gill nets set at depths of 72 to 84 feet in Lake Michigan, and eighty-five dead birds were caught in nets set at 240 feet near Wolfe Island, Lake Ontario, in May 1968.

No other duck nests in greater numbers in the high Arctic. The breeding range extends as far north as land is found and as far south as the tundra persists. According to aerial counts done by U.S. Fish and Wildlife Service biologists, the continental population is 3 to 4 million in early summer as nesting begins on the Canadian and Alaskan tundra.

To quote Arthur Cleveland Bent's *Life Histories of North American Wildfowl*, "Oldsquaws are lively, restless, happy-go-lucky little ducks during the winter, but early in the spring they gather in little flocks about some favored fe-male in fantastic postures, rushing, flying, quarreling and filling the air with their musical love notes. If noisy at other times, they are still more so now, vying with each other to make themselves seen and heard."

Arriving on the nesting grounds from mid-May to mid-June, they build nests in small cuplike hollows on the tundra, sometimes near pools but often at some distance from water. The nest is lined with mottled brown down from the female's breast, and the five to seven buffy olive eggs are covered when she leaves the nest to feed. During the twenty-four to twenty-nine days of incubation, the drake does not entirely desert her until the eggs begin to hatch.

The brown ducklings are blackish brown and white with no touch of yellow at all, quite unlike our Mallards. When they are about two days old the mother leads them to the water where they are taught immediately to dive.

Carl Kurtz

Bufflehead

Through March and most of April, I enjoy the delightful antics of the little butterball ducks, smallest of the divers. It weighs only ½ to 1½ pounds and is fat as a butterball in the fall.

Butterball is the Bufflehead duck. The obsolete word "buffle," meaning buffalo, refers to the handsome, big puffy head, black and glowing in the sunlight with iridescent green and purple. There is a large, white wedge-shaped patch on each side of the head. The neck, breast, and sides are snowy white, and the belly is light gray.

The back and wings are black, with white shoulders and speculum giving in flight a black-and-white striped effect. The eye and bill are dark. The feet are flesh-colored with the hind toe webbed, a sure sign of a diving duck—the puddle ducks have unwebbed hind toes.

The female is always smaller than the male, a subdued dark gray with a small white patch on the side of the head and a lighter gray breast and belly. The bill and feet are gray.

On a farm pond just north of Red Rock Dam we watched a small flock of five, two males and three females. This is one duck that does not need to resort to a sloping take-off from the water. The two males came straight up out of dives and into the air, flying low in formation in a long oval the length of the pond and back again; then lowering the pretty pink feet, they skipped on the surface of the water to a stop and into a dive. Time and again they performed this joyful act.

This was not the courtship display—there was no quarrelsome nature to this act, just joyful exercise. The courtship ritual is quite different. Then the male puffs his handsome head to twice the normal size, extends his neck with bill pointed up, and struts about as if standing on his tail, finally drawing his bill down on his puffed-up breast. Often the males fight with considerable viciousness.

Later on a Monday after school, my young student friend Phil Myers and I headed for Red Rock Bluff in the refuge to hunt the nest of a pair of Red-tailed Hawks. With water half a mile this side of the old heron rookery where the Redtail pair nested last year, there was no question of walking in. Taking the line of least resistance, we set up the 20-power telescopes on the edge of the gravel road, and seated comfortably in deck chairs, we searched a 2-square-mile area.

As often happens on a field trip, we were immediately distracted by a small flock of Buffleheads and Goldeneyes and completely forgot the hawks. The ducks were diving constantly, feeding on small fish, sometimes swimming just under the water so we could actually follow their pursuit of fish. We regretted that we had not brought along the stopwatch to time the dives.

Other observers give the time underwater as thirteen to twenty-three seconds, but we watched several dives lasting longer than twenty-three seconds by wristwatch count. Only once did two of the male Buffleheads rise straight up in the air, fly side by side the length of the pond, then ski on bright pink feet to a stop and into another dive.

Eventually we remembered the hawks. Searching the heron nests, we saw the head of a large bird above the rim of one nest. Hopefully it is the Redtail incubating eggs and not a Great Horned Owl. The herons can intimidate a Redtail, but the owl is not so easily intimidated. Woodward H. Brown, in "An Annotated List of the Birds of Iowa," rates the Buffleheads as uncommon migrants. Long ago they nested in northern Iowa in the lakes region, but most now nest in Canada. There may still be an occasional nesting in Iowa, as Jim Sieh banded a flightless immature specimen in Sac County, July 16, 1962.

The Bufflehead nests in a tree cavity 5 to 30 feet up, favoring deserted Flicker nest holes, returning to the same site year after year. The usual clutch is ten to twelve ivory-white eggs incubated by the female about twenty days. The downy duckling has rich brown upper parts and white underparts including the chin and throat. There is a large white spot on each wing and on each side of the rump.

Many of the diving ducks feeding at the top of the food chain, especially the Canvasbacks and Redheads, are endangered, due not to hunting, which is well regulated, but to the persistent pesticides, the chlorinated hydrocarbons. Dr. Larry Wing, research biologist, Iowa State University, reminds me that at least the entire species is not endangered by the new pesticides, whereas the persistent pesticides, by interfering with the reproductive process (thin-shelled eggs, etc.), endanger whole species.

It is the painstaking research of dedicated wildlife biologists like Dr. Wing and his graduate student, Ann Konermann, that will eventually save our wildlife from extinction. Dr. Wing directed comprehensive research in the Red Rock Refuge the spring and summer of 1973.

The Buffleheads' diet is mainly fish, and their flesh is consequently rank and fishy. To quote Jack Musgrove (*Waterfowl in Iowa*, 1943, illustrated by Maynard Reece, a gem of a little book), "Their small size and the quantity of animal matter they consume make them undesirable for the table. Considering these facts, together with their beauty and rarity, sportsmen should refrain from shooting them."

Mergansers

In 1967 twenty Common Mergansers, once our commonest cold-weather ducks, began diving for fish in the flooded areas of Red Rock Refuge on February 21.

The first forty Canadas also arrived that day and with them a beautiful Bald Eagle with snow-white head and tail. The first five hundred Snow and Blue geese came March 5, so the refuge is coming alive again—not that it was exactly devoid of birdlife during the winter. We had fifty-eight species within the refuge and three other species at the dam on our Christmas census.

The mergansers are ducks but not the usual flat, broad-billed types. Their bills are long and slender with toothed, serrated edges and a hook on the end ideal for holding slippery fish, hence the name they are also known by, Sawbill.

Beautiful divers, they pursue and catch fish under water. They want no part of mud dabbling, but they will fish in shallow water if it is teeming with schools of small fish.

Hooded Merganser

Red-breasted Merganser

Common Merganser

In flight they have a long, stretched-out look with head, neck, and body held in a straight, horizontal line. Although clumsy in leaving the water, they are rapid fliers, holding the line in formation, not bunching.

The male is a handsome bird with a long, white body, dark green head, black back, and red bill and feet. The delicate peach-bloom coloring of the breast is one of the loveliest colors in the bird world. The female is a subdued Quaker-gray with a crested rusty head. The male has a puffy head but no crest, and both have square white wing patches. One of our largest ducks, he is 20 to 27 inches long with a wingspread of over 3 feet.

The Wood Duck is not the only tree nester. This merganser prefers a hole quite high in a tree but will nest in holes under boulders near freshwater lakes. But don't look for a nest in the United States, as almost all of their nesting is done north of the border in Canada. In Europe they are called Goosanders, nesting in the far north there, even beyond the tree limit.

The nest of grasses, leaves, and moss is lined with soft white down from the breast of the female. The usual clutch is six to twelve eggs of a creamy buff color laid in late May and hatched in twenty-eight days.

The downy ducklings bounce to the ground, and they are little beauties with olive-brown backs and white-edged wings. The neck is pinkish, light cinnamon with snowy white stripes on the sides of the head.

Like most waterfowl, they have two types of migration, spring-fall and moult migration. In August, just before they shed their flight feathers and become flightless for a while, they fly to a safe lake out of reach of predators. This is called the moult migration.

The spring migration begins in late February, with the majority going through in March. Fall migration is usually late, beginning in November and lingering into winter.

They also have two types of plumage, the eclipse and the adult or breeding plumage. The eclipse plumage lasts at least from August to December or January—the male has gray body sides and a reddish brown head during this period. Full adult plumage is regained by January and lasts into summer.

Hunters do not prize the fishy-flavored flesh of the merganser—they are entirely fish eaters. Nevertheless, their ranks are thinning. Being at the top of the food chain, they eat fish carrying heavy loads of persistent pesticides—not just DDT but all the chlorinated hydrocarbons including dieldrin, aldrin, etc., effects of which are cumulative in the food chain and are known killers of wildlife.

The other two species of mergansers are rarer, the Red-breasted, an uncommon migrant, and the Hooded, an uncommon breeding bird, according to Woodward H. Brown's "An Annotated List of the Birds of Iowa."

The little Hooded Merganser is just 13 inches long with a wingspread of 26 inches. The male is one of the handsomest of the ducks, with black back, brown sides, white breast, and head topped by a big black-bordered white cockade. The female lacks the big cockade but does have a bushy reddish brown crested head, gray sides, and black back.

I once had the pleasure of watching three Hooded Mergansers diving for fish in a freshly stocked farm pond. They have voracious appetites and they were very hungry, eating small fish constantly during my hour there. I must confess I did not report the poaching mergansers on the posted pond to the farmowner.

However, Hooded Mergansers do eat a large amount of stems, leaves, and seeds of water plants besides fish. Their flesh has a better flavor, therefore, than that of the Common Merganser, but it is still fishy. They, too, nest in hollow trees, sometimes quite low. Their nest is composed of leaves and decayed wood well lined with down. The eggs number ten to twelve and require twenty-eight to thirty days incubation. The young bounce to the ground and are handsome with brown above, sides of head and neck pinkish buff, with throat and belly of pure white.

It is believed that a few still nest in Iowa. In 1960 Fred Kent observed a female and six young at Coralville. One older report (Dries and Hendrickson, 1952) was interesting. Two nests containing eggs of both the Hooded Merganser and the Wood Duck were destroyed by floodwaters at Lake Odessa, Louisa County.

Mergansers are generally silent, only occasionally making low croaking sounds quite unlike the incessant quack, quack of that loudmouth, the Mallard drake.

Larry Stone

Turkey Vulture

The annual watch to see the first Turkey Vulture at Red Rock begins in mid-March. This was true in 1980, but the first bird did not appear until March 27 in 1981.

The vultures' summer communal roosts were noted by the pioneers and are recorded in the early annals of the state. These traditional roosts in the valley of the Des Moines River have probably been used by the buzzards for hundreds and thousands of years.

When Lake Red Rock filled in March 1969 and for a few years thereafter, three roosts were occupied nightly from mid-March to mid-October. Now the two within the 26,500-acre refuge west of Mile-long Bridge have both been deserted except for occasional use by fall migrants.

The buzzard was one of the birds of my childhood. The now-deserted Red Rock Bluff roost was only 3 miles northeast of our farm home. Every morning the big birds soared over the farm as they began their search for carrion, and every evening they soared north again to roost.

Our only active roost now is the one on top of Elk Rock Bluff in Elk Rock State Park on the south side of the lake. It had thirty birds all summer. These are mostly birds under the breeding age of three years.

Lake Ahquabi State Park near Indianola had a roost of eighty birds the summer of 1980; Lacey-Keosauqua State Park in Van Buren County had forty; Ledges State Park near Boone, twenty-four; and in Madison County, Gene Armstrong reported a roost of twenty birds in the Middle River valley and a roost of fifty in September in the Raccoon River valley, both in privately owned timber.

The vulture is now a blue-listed or threatened species. Its decline is attributed to pesticides and pollutants, loss of habitat, and a shortage of carrion as a result of improved veterinary care of domestic animals.

These big birds require a secret, safe place to nest. In 1980, Elk Rock State Park rangers on April 27 found one egg in a nest in an old building in a remote area. The nest was depredated, and the parents may have been the same pair that nested later in a little cave in one of the sandstone bluffs in the park. This cave has been used almost every year since the lake filled, except during high floods.

Among my slides is a series of the nesting and rearing of baby buzzards taken by Iowa State University graduate student Joseph Schaefer. The female is often reluctant to abandon her eggs, and one slide shows one standing over her eggs with her head down, ostrichlike, and her back turned to the camera. The tiny babies are covered with long, snow-white, fleecy down that persists even after the big primary wing feathers are fully grown. On one slide, the naked black bellies of the young, stuffed with carrion, protrude from the white fleece, a ridiculous sight that always brings laughter from the audience when I show it. The incubation period is forty days,

and the young remain in the nest about two and a half months, fed by their devoted parents.

In all my years of caring for young or injured birds, I had never received a buzzard until September 22, 1981, when Jeff Emal of rural Pleasantville rescued one and brought it to me. It was weak and listless and unable to fly. I took it to my veterinarian, Dr. Mark Poell, the next morning. He found no injury; the bird had not been shot. His tentative diagnosis was possible pesticide poisoning, and he flatly stated that the bird would probably not recover. Nevertheless, kind soul that he is, he gave me antibiotics, vitamins, and other medicines to give twice daily.

Now there's just one way to get medication and food down a buzzard's throat and that, unpleasant as it may seem, is to poke it down with thumb and finger and then still farther down with the forefinger.

"Buzzy" spent all its time the first two weeks on the ground in one corner of its flight cage. When I entered the cage, it played "ostrich," putting its head on the ground with its back turned toward me.

Self-defense in a cornered vulture means regurgitating rotten food all over the enemy. Sometimes Buzzy would upchuck the food and medicine, and then I had to poke it all down his throat again, a messy procedure, to say the least. And I learned to dodge, too.

Frankly, I had never seen a vulture's head up-close before. The beak is pinkish and huge, not shaped like any other beak I had ever seen, long and strong and hooked at the end. The eyes are dark. There are small, white, wartlike growths in front and under each eye. The top of the head has a series of folds like accordion pleats. The dark skin is almost naked except for a few stiff, short bristles.

The plumage is beautiful, the feathers dull black with a brown edge. The upper surface of the great wings is black, but the underside of the primary and secondary wing feathers is a lovely silvery gray. The feet and legs are grayish brown, with ordinary nails, not talons.

Buzzy was a big bird, 2 feet long with a 6-foot wingspan. I would judge it was a juvenile hatched in late May or mid-June.

For two weeks Buzzy showed no signs of recovery. It remained listless and standing on the ground. On October 9, I noticed the first improvement. It was perched on the 3-foot pedestal in its cage. And on that day, Jeff brought half a dead pig, about 20 pounds, and put it in the cage. And I decided that Buzzy could feed itself. And this it began to do, eating a little more each day, a really hopeful sign. The riper the pig, the more Buzzy ate.

On October 14, Buzzy was perched on the high perch pole. It could fly and was well on the road to complete recovery.

I phoned *Des Moines Tribune* photographer George Ceolla, asking him to photograph the release on Elk Rock Bluff near the roost. George came on October 19, and Buzzy and I rode with him to the bluff. George took close-up shots of the very ugly, utilitarian head and then, using the telephoto lens, more shots as Buzzy became airborne and soared around the west side of the wooded bluff.

We had released it just in time. By the end of the next week, the roost was deserted. Buzzy and its tribe had migrated south.

On every field trip between late March and mid-October since 1967 I have watched with pleasure the black Turkey Vultures sailing gracefully, one by one, to their roost trees in the Red Rock wildlife refuge. I have watched

them sunning on sandbars, so ridiculously clumsy walking in contrast to their grace in flight. With a 6-foot wingspread, angled slightly upward to form a shallow V or dihedral, they can soar for miles with little effort—just a few flaps now and then to catch the next updraft.

The Red Rock area has the largest vulture population in the state, with a total of 150 to 160 of the big birds occupying three roosts this past summer—one just west of Red Rock Bluff near the Great Blue Heron rookery, one north of Old Pinchey, also in the refuge, and a third on south Elk Rock Bluff in Elk Rock State Park.

On September 1, 1972, Stuart and Eunice Kuyper, who have a beautiful home on Lake Red Rock, invited two Central College professors and me to dinner and a cruise on the lake. As we came alongside Elk Rock Bluff I said, "I do wish we could get those vultures up and get an accurate count." No sooner said than done! Stu picked up a beautiful, slender, funnel-shaped brass horn not over 14 inches long—a canalboat horn acquired in Holland—and blew a few blasts that would wake the dead. Every bird rose straight up into the air and then soared gracefully around and around above the bluff. We counted 64. A very pleasant count method, indeed.

During September and October, as migrating flocks from the north move in to spend the night, the roost at Red Rock Bluff sometimes numbers over 200. The resident flock rises as if to greet the migrants, and there is a milling, soaring powwow high in the air often lasting an hour or more. Occasionally one darts at another but makes no contact—they are peace-loving birds.

These are traditional roosts probably used for centuries past. They were noted by the pioneers 125 years ago. And the first buzzards I ever saw, over 60 years ago, were sailing north over my home 3 miles southwest of Red Rock Bluff.

Why are there so many buzzards in this area during the summer breeding season? Many are nonbreeders; like so many other large species—eagles, hawks, herons—they do not breed until three to four years old.

Do they locate dead animals by sight—they have marvelously keen eyesight—or by odor? That remains a violently disputed question. The answer, I suspect, is both. What to us is a stench is to them an alluring Chanel No. 5.

They are seldom seen in great numbers on nearby farms because they fan out as far as 50 miles in the search for food. Although classed as birds of prey, they are not killers. They eat only carrion, any species—snakes, including cottonmouth moccasins (except the head and poison fangs), alligators, fish, birds, and mammals, including skunks (they skip the scent glands).

They do not always feed every day, perhaps not for several days. Dr. Daniel E. Hatch, University of Nebraska, in his comprehensive study of the Turkey Vulture, found them to be very hardy birds, able to go without food and water for eight to eleven days and recover without ill effects.

In 1969 the buzzards in the Red Rock Bluff summer roost numbered sixty-three. During August I kept a daily record of the number returning each afternoon. Fewer than 50 percent and often as few as 10 percent had been out hunting each day.

In pioneer days they were valued as sanitarians, consuming all dead livestock. (No, they do not spread hog cholera or anthrax; tests in Florida in 1914 proved that their powerful digestive juices killed all disease organisms.) Now Iowa requires immediate removal of dead livestock, and the big birds sometimes are hard-pressed for food.

I once suggested that road-kill deer be placed in a remote area of the refuge for the buzzards and Bald Eagles, too. Needless to say, that suggestion fell like a lead balloon, and I was informed that the law prohibited it—a stupid law if no reasonable exceptions are permitted, especially in view of the population decline of the vultures in recent years. Now on the tentative Audubon Society's Blue List, this means that further decline will place them on the endangered species Red List.

Pesticide kills (especially fish kills) are of course eaten here. On their winter grounds in the southern states they are still endangered by mirex, a very deadly poison needlessly and senselessly used on millions of acres to control the fire ant.

Such a bird is limited in the number of places it can hide its nest and be safe from predators. Caves, ledges, cliffs, hollow stumps, hollow logs on the ground, brush piles, abandoned barns—all have been recorded as nest sites in Bent's *Life Histories of North American Birds of Prey*.

The two or three creamy-white eggs splotched with brown are incubated about thirty days. On hatching, the young vulture is covered with long, white, cottony down. At eight weeks the wings are fully developed, the back and sides of the breast well feathered, but down remains on the neck and underparts. They are unable to fly until ten to twelve weeks old. The naked head of the young is black, unlike the bright red head of the adult.

Many of the farmers along the south border of the refuge are lifelong residents, descendants of the pioneers, and many are keen observers. One of these young farmers, Ralph Dyer, roamed the wooded bluffs as a boy and acquired a variety of wildlife pets. Once he found a nest in a hollow log and brought home a big dark nestling, a young buzzard. He fed it, cared for it, and tamed it. The affection was mutual, and the young buzzard followed Ralph like a young puppy.

His mother, Almeda Dyer, recalls with great glee the shocked surprise on visitors' faces when they saw the small boy with the big ungainly buzzard waddling along at his heels.

Carl Kurtz

Osprey

The presence of a pair of Ospreys, our only fish hawk, has been a special pleasure this spring of 1978 in Red Rock Refuge.

On Saturday, April 1, the pair perched in an old dead tree on the floodplain just east of refuge headquarters, digesting their fish breakfasts and occasionally preening. They paid no heed to the 10,000 Snow, "Blue," and Canada geese putting on an aerial show to the north, nor did they notice the flocks of ducks on the water below.

On Saturday, April 8, while touring with members of the Marion County Teachers Conservation Council, we watched this graceful hawk fishing below Painted Rocks Bluff, a magnificent setting and a magnificent bird. Last fall one arrived at the lake on Labor Day and fished the waters until November before continuing its journey south.

A winter sighting in Iowa is extremely rare, but Jon Stravers observed three in January 1977 fishing the Mallard's open water-hole just west of the dam. A lazy Bald Eagle harassed them, stealing their fish despite the angry screaming frenzy of the Ospreys.

An endangered species, they are recovering slightly since the banning of DDT and dieldrin in the United States, but they are not safe in Mexico or Central America where DDT is still widely used.

It is the most cosmopolitan of all our raptors, found in most temperate and tropical regions of the world. A big hawk, 22 inches long with a 54-inch wingspread, it is dark above and white below, with a conspicuous crook in the long wings and black "wrists" at the bend of the wings. A black line through the eye and a white line above are not always visible in flight.

Despite their widespread distribution in North America, the breeding sites are limited to good fishing areas. In southern Massachusetts and Rhode Island, they were welcomed on farms and estates as picturesque features or because they were good watchdogs, giving shrill angry screams at the approach of other hawks or intruders.

Their huge stick-nests were built in the tops of trees. Often the landowner set a tall pole with a cartwheel on top, which the Ospreys readily accepted as a nest site.

The three pinkish white eggs, heavily

blotched with reddish brown, are the most beautiful of all hawks' eggs. The female incubates the eggs twenty-eight days and is fed by the male except for short periods of exercise or fishing. The young, fed by both parents, remain in the nest two months.

The Osprey almost only eats fish, though it is known to take other items. Flying along, intently scanning the water, flapping or sailing, occasionally hovering over a spot, it plunges down, hitting the water breast-first with wings pointed upward, sending the spray flying. It seizes the fish in its powerful "fishhook" talons. I have seen it catch a foot-long catfish, and others have reported 4- to 6-pound catches.

Larry Stone

Bald Eagle

Winter at Red Rock Refuge is exciting. Besides thousands of geese and ducks and a couple hundred Ring-billed Gulls, there were fifteen Bald Eagles in December of 1969.

Thirteen of the eagles were immatures and two were adults. The immatures migrate earliest; the first one arrived October 16 and by October 28 there were twelve. One adult arrived November 6 but moved on. Two more adults arrived November 22.

I have spent many hours trying to determine the age of the immature eagles. According to Dr. William E. Southern, Northern Illinois University, an authority on the wintering Bald Eagles along the Mississippi River, the plumage varies each year through the sixth year, when the big bird finally acquires the snow-white head and tail plumage of the adult.

Only one of our adults had a snow-white head and tail. The other adult had a creamy head and tail, fifth-year plumage. The white feathers were topped with brown, giving a creamy appearance.

Three had fourth-year plumage, sides of head and throat white, some white in tail, breast brown, and beak yellow.

One had the distinctive third-year plumage with a whitish lower breast and belly and a triangular patch of white feathers on the back between the wings. (Not all third-year birds have the triangle of white.)

Two had belly and lower breast of light tawny-brown with the upper breast darker like a dark bib, also some white in tail, second-year plumage.

The first-year bird is mainly brown with some white in tail. I am sorry to say that not one seems to fit this description.

According to the scientists, the persistent pesticides present in fish and game eaten by the eagles cause adverse behavioral changes and metabolic disorders such as thin-shelled eggs that break during incubation. The adult eagle is killed by a DDT-derivative accumulation in the brain of only 56 to 86 parts per million. An endangered species is a red warning sign of a dangerously contaminated environment.

Observing the behavior of the fifteen Bald Eagles at Red Rock Refuge was a real treat for me. In past years eagles in central Iowa were a rarity, although they winter along the Mississippi and Missouri rivers.

The study of animal behavior is called ethology. What I have observed would probably be classed as maintenance activities by the ethologist, but I am not a scientist; I am strictly an amateur.

One day two eagles indulged in aerial combat. One flew after the other. When the pursuer was directly overhead, the lower one rolled completely over on its back and struck upward with its talons but missed.

Seven were congregated on the stumps and snags along a narrow slough. One on a 2-foot stump peered intently down into the water when another eagle rudely knocked it off its perch and into the water. It did not retaliate. Instead it took a leisurely bath.

One afternoon four of them took great splashing baths in the wheat field pothole pond. They spent at least fifteen minutes showering water 6 to 8 feet in the air. The afterbath ritual of body shaking, wings partly outspread, and much preening occupied fully an hour.

During this, about fifty Ring-billed Gulls circled over the pond, occasionally dipping down to catch small fish. Neither gulls nor eagles were disturbed by the presence of the other. A Great Blue Heron reacted otherwise. It flew in, stood with long neck stretched up watching, then departed, although this is one of its favorite fishing holes.

On another occasion I watched one eagle fly across the wheat very low when it suddenly dropped down and struck something small with its talons. I speculated that it was a white-footed mouse, but the eagle continued to jump up about a foot in the air and pounce down on it with its talons. It did this at least ten times before I was able to see that its prey was a small beverage or fruit-juice can. It pounced about ten or twelve more times rather like a playful puppy before it continued its low flight across the field.

The late Charles Broley, eagle man of Florida who had handled over 1,000 young eagles in the nest, found them to be great curio collectors. He found light bulbs, Clorox bottles, clothespins, shells, and many other oddities in their great nests. One female incubated a white rubber ball for six weeks after her two young hatched.

Bald Eagles prefer fish to all other food, but they will take other game. One day Linda Beamer observed twelve immatures kill a goose on the wheat and eat it on the spot. (They cannot lift more than 6½ to 7½ pounds.) They fight furiously like roosters, jumping 3 and 4 feet in the air and striking at one another with razor-sharp talons.

I presume there is a dominance pattern but I do not know. One day two birds fought for a couple of minutes until the vanquished one retreated from the pothole congregation to stand all alone some 200 feet away on the wheat. Rejected and dejected?

They are magnificent soaring birds, but their walking is not the most graceful. It can only be described as a clumsy, lumbering gait. Surprisingly they walk quite well on the ice.

Dry oak woodland: white oaks, basswoods, and bracken fern on upland along the Iowa River, Iowa County. Photo by Carl Kurtz.

Red-eyed Vireo at its nest in a silver maple tree. Photo by Carl Kurtz.

Male Downy Woodpecker on a tree knot. Photo by Linda Kurtz.

Male and female Rose-breasted Grosbeaks at their nest in an apple tree. Photo by Carl Kurtz.

Wet prairie: sword grass, sedges, brook lobelia, and purple gerardia at Silver Lake Fen, Dickinson County. Photo by Carl Kurtz.

Sora feeding on the edge of a small pond. Photo by Carl Kurtz.

Male Common Yellow-throat at his nest with a cowbird young. Photo by Carl Kurtz.

Sedge Wren singing on a compass plant. Photo by Carl Kurtz.

Rich forest: mature northern red oaks with may-apples and lady ferns at Dows State Preserve, Linn County. Photo by Carl Kurtz.

Male Northern Oriole carrying food near his nest. Photo by Carl Kurtz.

Male Red-bellied Wood-
pecker on a suet post. Photo
by Carl Kurtz.

Male American Redstart
in spring migration. Photo
by Linda Kurtz.

Marsh: cattails along the marshy edge of Big Wall Lake, a prairie pothole marsh, Wright County. Photo by Carl Kurtz.

Virginia Rail feeding in shallow water. Photo by Carl Kurtz.

American Coot on the edge of a muskrat house in a cattail marsh. Photo by Carl Kurtz.

Yellow-headed Blackbird perched on bulrushes in a prairie marsh. Photo by Carl Kurtz.

Open water: fragrant water-lily in the back-waters of the Mississippi River, Clayton County. Photo by Carl Kurtz.

American White Pelicans in open water at Lake Red Rock. Photo by Carl Kurtz.

Great Egret surrounded by duckweed-covered water. Photo by Linda Kurtz.

Ring-billed Gull flying over the clear water of West Lake Okoboji. Photo by Linda Kurtz.

Tallgrass prairie: dotted blazingstar and big bluestem at sundown on Cayler Prairie, Dickinson County. Photo by Carl Kurtz.

Male Bobolink at his nest with young. Photo by Carl Kurtz.

Savannah Sparrow on western yarrow with food for its young. Photo by Carl Kurtz.

Upland Sandpiper perched on a fence post. Photo by Linda Kurtz.

Savanna: shooting-stars, downy phlox, hoary puccoon, and wild columbine beneath an old open-grown bur oak at Rochester Cemetery, Cedar County. Photo by Carl Kurtz.

Red-headed Woodpecker at its nesting hole in a dead American elm. Photo by Carl Kurtz.

Male Eastern Bluebird at his nest hole in an old fence post. Photo by Carl Kurtz.

Male Indigo Bunting with a mulberry at his nest. Photo by Carl Kurtz.

Floodplain forest: old silver maples in standing water covered with duckweed on the Mississippi River, Allamakee County. Photo by Carl Kurtz.

Male Prothonotary Warbler in backwater habitat. Photo by Jim Messina/ Prairie Wings.

Male Wood Duck in woodland habitat. Photo by Linda Kurtz.

Male Yellow-bellied Sapsucker feeding on sap flowing from a sugar maple. Photo by Carl Kurtz.

Carl Kurtz

Marsh Hawk

Red Rock Refuge has its share of the birds of prey, both hawks and owls. The greatest numbers are seen during spring and fall migration, but even the 1970 Christmas Bird Count had five species of owls and seven species of hawks, with a high of twenty-four Red-tailed Hawks, five Marsh Hawks, and four Bald Eagles.

The Marsh Hawk, 17 inches long, is America's only harrier. The birds are slim, with long, slightly angled wings and long tails. Their flight is usually very low and buoyant, with wings held in a shallow V rather like the up-tilted wings of the Turkey Vulture.

The young birds, brown-and-gold beauties with white rump patches, migrate first each fall, usually appearing in September, while the adults come through later. The female is brown-backed with a gold breast, streaked with brown. The male is slightly smaller with a gray back and white breast. Both have white rump patches.

On many days, while Bald Eagles soared ever higher overhead, I have watched the long, slender, graceful Marsh Hawk "quartering" the valley floor below. Quartering is a hunting term meaning to pass back and forth across a field in many directions.

Let not this brown-and-gold beauty deceive you, this bird is a highly skilled killer, but it kills only to eat, one of the predators in nature's master plan. Its favorite diet is mice but it is not a bit above taking birds, sometimes hovering over a bush five to ten minutes until it routs the bird and kills it. The prey is usually eaten on the ground, but one day as John Beamer, refuge manager, watched, a Marsh Hawk dropped down, picked up a mouse in her talons, then, apparently frightened by John, she flew, reached down with her beak, removed the mouse from her talons, and ate it in mid-air.

A flood-control lake is often very wasteful of wildlife. When the refuge flooded in late February, all the mammals—mice, voles, rabbits, etc.—were driven out or drowned, but what is disaster to one species is gourmet dining to another. On this occasion John watched three Marsh Hawks patrolling the edge of the floodwater and picking off the fleeing mice, just as in pioneer days observers reported the Marsh Hawks hovering in front of roaring prairie fires catching the fleeing mice.

In the long run, not much is wasted. As the water recedes, the egrets and herons move in to fish the little pools. Bald Eagles take stranded fish, and the Turkey Vultures leave only the bare bones of dead mammals and fish. Later, the shorebirds, sandpipers, and plovers clean the mud flats of insects and other small creatures. In late summer of 1969, I found thousands of tiny mussel shells picked clean of meat. Who was the gourmet diner? I never knew.

Marsh Hawk courtship is no humdrum affair; it is quite spectacular. The nuptial flight of the male is a series of nosedives like capital U's strung together, each dive about 50 feet in depth. Sometimes in the excitement the female joins in the diving too, according to the late C. L. Broley, Bald Eagle bander.

Unlike the other big hawks that prefer tall nesting trees, this hawk chooses to nest on the ground in a marsh or an upland meadow. Although the number of breeding birds has decreased markedly in recent years and they now rarely ever nest in Iowa (Woodward H. Brown, "An Annotated List of the Birds of Iowa"), we have at least one comprehensive nesting study (reported in *Iowa Bird Life*) done in recent years by Dr. Robert Vane of Cedar Rapids and Fred Kent of Iowa City, both expert birders and skilled photographers.

They discovered the nest June 9 in a depression in the ground, containing five white eggs, in a waving sea of marsh grasses in Johnson County. "A photographic blind was placed approximately 40 feet from the nest on the evening of June 21, and after leaving the area the female was seen to return to the nest within a very short time," their account says. "Sunday, June 23, dawned bright and clear with the two in the blind at 5:30 A.M. Two young hawks and two eggs were present in the nest." (The twenty-eight-day incubation time begins with the laying of the first egg; consequently, the youngest are often unable to compete for food.)

"Within minutes after entering the blind the female bird had returned to the nest with a mouse to feed the young. The prey was held with one foot, while a piece of bloody meat was torn away with the beak and fed to first one and then the other of the young hawks.

"The female was the only one seen at the nest itself, whether feeding young or brooding, but the smaller male was in the area a great deal of the time flying nearly perfect figure-eights directly over the blind and call-ing his monotonous 'kikiki.' Interestingly enough, the male bird was a young cock and had not assumed the gray plumage as yet, being brown like the female, but a lighter brown and definitely smaller.

"Frequently the female, on returning to the nest, would bring a beakful of dried grasses which she would apparently drop at random after alighting to brood the young. If the young should push themselves off into the grasses at this stage, the female, using her beak, would pick them up by the neck and place them beneath her in the nest."

On June 30, there were three young in the nest and one sterile egg. And by July 7, the young were seen in tunnels of grass where they were avoiding the hot sun. At intervals they would emerge from their tunnels to flap their wings in the open nest area. On approaching, the young assumed a defensive attitude, beaks open and talons up. On July 20, they were quite fully feathered, showing the white tail coverts, and were taking short flights. The movies made of the nesting study are indeed very beautiful.

On July 1, 1970, for the first time in the history of Iowa a law prohibiting the shooting or killing of any bird of prey went into effect, thanks to the efforts of Wesley A. Newcomb, U.S. Game Management agent, and an enlightened legislature. However, the drastic population decline of all birds of prey in the early 1970s was not due to hunters but to the widespread use of the biocides—the persistent pesticides, DDT, dieldrin, aldrin, etc.

The life of a species is about 500,000 years, according to Pierce Brodkorb, with the loss of no more than two species per century. The loss of eighty species over the last 300 years is fifteen times the normal evolutionary rate.

F. K. Schleicher/Vireo

Cooper's Hawk

On May 6, 1983, while doing the "birdathon" count to raise money for the rehabilitation of injured or orphaned hawks, owls, and songbirds, Ann Johnson and Jim Sinclair of the Rolling Hills Audubon Society located a Cooper's Hawk nest in Robert's Creek County Park on the north side of Lake Red Rock, the first documented nest record in Marion County in twenty-seven years.

The Cooper's, 15 inches long with a 28-inch wingspread, is one of three accipiters that reach Iowa. Long-tailed, long-legged, with short rounded wings, they fly rapidly through thick cover, feeding mainly on birds and small mammals.

The female, larger than the male, has fine cross-barring on the breast and belly, a dark slate-blue back, barred tail, dark cap on the head, and fiery red eyes. It is often called the Little Blue Hen Hawk.

The bird was incubating, her long tail protruding over the edge of the fairly large, flat stick-nest 20 feet up in a honey locust tree entwined in Virginia creeper vines. Unfortunately, the nest could not have been located in a more dangerous place—at the west end of a very busy campground and only a few feet away from a motorbike trail through the beautiful forest. Shy and secretive, these birds do not react well to human intrusion, and we feared the nest would be abandoned long before the twenty-six-day incubation period ended.

Jon Stravers of Pella, doing research in northeast Iowa on the breeding biology of Red-shouldered Hawks, was home on a weekend and talked to park rangers and the executive director of the Marion County Conservation Board about the possibility of closing the noisy motorbike trail. He explained that this hawk is blue-listed, suffering a serious population decline, and in fact has been placed on the state endangered species list in Missouri. The trail, however, was not closed.

So we worried, watched, and hoped. The brave bird stuck to her nest, her tail visible every time I visited the area. Then one evening, with campers only 25 feet from the nest, she was absent and did not return even by sunset. Sadly, I returned home, sure she had deserted.

Ever hopeful, I made another visit a few days later, and she was sitting tight. I suspected she was brooding tiny, white, downy hawklets. Stravers later confirmed this—three babies in the nest and one dead at the base of the tree.

Now I was no longer worried. The parental instinct is too strong for desertion at this period. A Great Horned Owl would be the only predator brave enough to attack the parents.

On the morning of June 22 I met Stravers at the campground. After donning long spikes, he climbed the tree and banded the oldest hawklet with a numbered U.S. Fish and Wildlife band while the mother flew in wide circles and cackled.

The two younger hawklets were too small to band. Covered with thick, woolly, cream-colored down and emerging feathers, they were handsome birds. How I wished I could hold one of them.

A week later Stravers returned and banded the younger two and reported the older one was out of the nest. They usually leave the nest before the flight feathers are fully grown and hop from branch to branch while still being fed by the parents.

By mid-July the nest was empty, and we were happy. Now the parent birds would spend the rest of the summer teaching the three to hunt.

Most of our eagles, hawks, and owls are declining in population, either on the blue or red list. The three major causes of the decline are loss of habitat, pesticides, and illegal shooting, in that order.

Gaige Wunder

Goshawk

Major invasions of northern birds in Iowa have been almost nil this winter of 1982–83 with but one exception. The Goshawk, also known as the Big Blue Darter, staged a major invasion.

It began in September with one record. By the end of November, Goshawks had been reported in Allamakee, Bremer, Clarke, Delaware, Iowa, Johnson, Marion, Polk, Story, Warren, and Wayne counties.

The Goshawk is the largest of the accipiters, the bird hawks, noted for their short rounded wings and long tails. Longer than our Redtail and more slender, the Goshawk is a handsome bird with a steel gray back, a black cap, a dark patch on the side of the head, and a white line over the bright orange-red eye. The breast and belly are silver gray finely marked with narrow, dark streaks.

Woodward H. Brown, in his "An Annotated List of the Birds of Iowa," stated that from 1945 through 1970 all records of Goshawks were from the northern half of the state. However, in recent years a few have been seen every winter in southern Iowa.

Why is there this extension of the winter range? Have the intensive row crop farming and habitat destruction suppressed the prey animals, or is it because of unknown factors?

This "king of winter" may stay in the boreal forest breeding grounds in Alaska and Canada all winter but usually drifts down to southern Canada and the northern states.

Recent invasions indicate a population decline when compared to those of the late nineteenth and early twentieth centuries. (The other two accipiters, Sharp-shinned and Cooper's, already are threatened species.)

Dr. Thomas Roberts, in *Birds of Minnesota*, mentions the heavy flights of 1917–18, 1920–21, and 1926–27 extending across the northern states from coast to coast. Naturally, they were attracted to the state game farm at Lake Minnetonka where 150 were trapped or shot. (All hawks and owls in Iowa are protected by law.)

The female Goshawk is larger than the male and is a fierce protector of the nest, eggs, and young. Our state nongame biologist was doing field studies on grouse in northern Michigan when he inadvertently got between two young Goshawks and their mother. Suddenly, he realized that a gray flying-tiger was about to attack. He hit the ground and she missed, but she quickly circled for another attack. He saved himself by rolling under some evergreen branches.

In early December our Red Rock biologist and two companions were duck hunting, well hidden in a camouflaged boat with plastic decoys nearby. A Goshawk came down, struck one decoy, circled, and returned to strike a second decoy and then a third decoy before flying off (and they didn't have a camera).

On January 7 Robert Thornberg of Pleasantville was observing a Long-eared Owl and two Sharp-shinned Hawks in a pine grove on the north side of Lake Red Rock when he glimpsed a big gray bird flying straight toward his head. He ducked. The Goshawk swerved but was so close Thornberg could see the fiery red eye.

At 5:30 P.M. on March 1, Jim Sinclair of Indianola and I were standing in the Conservation Commission parking lot on the bluff south of Runnells when an adult Goshawk flew east just below the level of the bluff about 25 feet from us. The bird flew straight through the willow thicket. We knew when it reached the bay beyond because a cloud of gulls exploded into the air. The next evening at about the same time the Goshawk flew west as we stood on the bluff.

Carl Kurtz

Broad-winged Hawk

In my first story printed in the *Des Moines Sunday Register* on October 5, 1969, I told of the thrilling experience of seeing a kettle of Broad-winged Hawks. They came up out of the timber on the north side of the Red Rock Refuge just west of the Mile-long Bridge.

They circled around and around like a huge chimney of a boiling kettle, rising ever higher until they barely could be seen through binoculars. Only then did they head south.

Occasionally, over the intervening years, we have seen a kettle of hawks above North Elk Rock State Park—twice during Outdoor Days for Knoxville's East Elementary third-graders, a thrilling sight for all. To avoid long waits in line, we used three high-powered spotting scopes. The children took turns looking through the scopes. They saw White Pelicans with 9-foot wingspans, the biggest birds to reach Iowa, Great Blue Herons standing 3 feet tall on stilt legs, big, black Double-crested Cormorants, Ring-billed Gulls, and others.

Because we study both insect and bird migrations, I was very busy tagging a monarch butterfly, showing the students how to identify the male by the black dot on a vein of the hind wing, and finally feeding the monarch sugar water. My attention was directed to the sky at the north end of the Mile-long Bridge. There a large kettle of Broadwings, probably seventy-five, was circling around and around, soaring ever higher in the updraft of a column of warm air. That big kettle was followed by two smaller ones, rising up to the bottom of a big, puffy white cloud. The children were seeing a truly magnificent migration, something to which no book can do justice.

A kettle is a spiral column of hawks. They have found a column of air, either a mechanical updraft caused by the vertical deflecting of the wind against a sandstone cliff or a thermal updraft, which is the rising of warmer, less dense air through cooler and denser air.

These small, chunky buteo hawks, 14 to 16 inches long, nest in the forests of our northern states and Canada and are often referred to as "birds of the wilderness." Their diet consists of frogs, toads, snakes, mice, crayfish, red squirrels, and a very few birds. During their migration to the tropical winter grounds, they often feed on big katydids in Panama.

Our Broadwings go through the Duluth funnel flying area at the west end of Lake Superior—they do not like to fly over a big body of water. The migration through the funnel is one of the great sights of the bird world. Dr. Pershing B. Hofslund of the University of Minnesota at Duluth heads the fall hawk count there. Often, 60,000 hawks pass through in September and October.

Broadwings migrate in mid-September until October 1. Young Sharp-shinned Hawks also migrate at this time, but the adults come through in mid-October.

Carl Kurtz

Swainson's Hawk

The Swainson's Hawk is a big handsome buteo hawk, the size of a Redtail with a brown head, back, and wings, a white throat and belly, and a finely barred tail. The best identification mark is the wide dark brown band across the chest. (The Rough-legged Hawk has a dark brown band across the belly and a black band across the end of the white tail.)

Swainson's is a western bird and fairly rare in Iowa. Woodward H. Brown, in "An Annotated List of the Birds of Iowa," stated, "Observations have been made as early as March 23 and as late as Oct. 31. There is also a winter record of a pair in Polk County in

1961-'62." It is, however, generally conceded that the Swainson's is absent from Iowa during the winter.

I was surprised to observe one soaring over Red Rock Refuge headquarters on March 1, flying a course from southeast to northwest and accompanied by two Marsh Hawks. One was playing, darting down at the back of the Swainson's but never quite touching it. The dignified Swainson's simply ignored the harassment. This behavior by a Marsh Hawk was new to me, but several times I had observed a kestrel playfully harassing a Redtail perched on a fencepost. A highly migratory species, the Swainson's winters mainly in Argentina. Bent's *Life Histories* records the great migratory flocks returning in spring. One observer in April 1890 watched a flock of 2,000 alight on his ranch along the Powder River in Montana. Many of the cottonwoods held 50 hawks. Others on the ground among the cattle appeared to be very tired and sleeping.

Ludlow Griscom in 1932 referred to the great flocks that pass through Central America as "one of the sights of the bird world." The great majority passed over the area in just a few days during October; a few enormous flocks took hours to pass a given point.

This species is now blue-listed. The breeding population in the southern Great Plains region and in Kansas is clearly down and in the mountain West region has declined or is absent. Its status is reported as stable in the Southwest region.

None of the biocides are banned in Central and South America. Only DDT and dieldrin are banned here. Other persistent pesticides were sprayed over thousands of acres in the West last year, resulting in known contamination of waterfowl. The ban on 1040, a deadly poison, has been lifted.

The Swainson's is considered totally beneficial because of the huge numbers of grasshoppers and crickets in its diet. In a stomach analysis study, one bird contained 86 grasshoppers, another 96, and a third 106.

Swainson's Hawks have been observed catching large insects on the wing, "volplaning, somersaulting, circling, then thrusting out a foot to catch the insect." They also take many rodents—mice, rats, ground squirrels, rabbits—and snakes, including rattlesnakes. One California observer reported a dozen of these hawks near dusk perching on gopher mounds waiting to snatch the emerging gophers.

They prefer open grassy fields with a few trees along streams for nesting and perching. The big, flat stick-nest is often 3 to 4 feet across and 20 to 50 feet off the ground. Incubation of the two, rarely three or four blue-white eggs faintly marked with brown requires twenty-eight days. The duty is shared by both sexes. The nestlings remain in the nest four to five weeks.

State Ecologist Dr. Dean Roosa studied several nests in north-central Iowa several years ago, but most of that habitat has since been destroyed. I had only one breeding record in Iowa in 1981.

Because of two adult color phases, the light and the rarer dark phase—also the streaked breast and belly of the immature—the Swainson's is often difficult to identify. Therefore, all identifying marks should be carefully noted, if possible, by birders.

Red-tailed Hawk

The Red-tailed Hawk is the only large hawk still nesting in Marion County, a permanent resident, although migrants from Canada move through this area in October.

Jon Stravers of Pella, a hawk authority, works much of his time researching Redtails and other birds of prey. In February and March he searches the tall timber for the big 3-foot-wide stick-nests, later observing and recording data on incubation and nestlings. Finally donning long spikes, he climbs the trees and bands the nestlings. In the spring of 1978 he located fifteen nests and banded twenty-three nestlings. Three nests were failures.

One of these nests was located in a ravine along Highway 14. I proudly showed a tour group of Marion County teachers the nest one May. It contained two handsome half-grown hawklets which were covered with thick, puffy, feltlike down.

In the hours after midnight on May 13, tragedy struck this nest when 50 mph winds blew it down, killing both young. That morn-ing Jon made the round of nests, sick at heart. At one three-bird nest he found the nest down and two babies dead, but the smallest one was still alive though weak. He and wild-life photographer Carl Kurtz brought the hungry hawklet to me, and we fed it an emergency meal of hamburger. (Young hawks and owls will die of rickets if kept on an exclusive beef diet.) Jon kept the baby at his home a couple of days, feeding it on natural foods such as mice.

He then took the baby to a nest containing only one much-larger hawklet, climbed the tree, and placed it in the nest. To his joy the parent Redtails adopted the little bird. The foster parents' original offspring fledged successfully within ten days of the transplant.

Two and a half weeks after the first trans-plant, Jon introduced into this nest a second Redtail nestling, a "brancher," meaning that it was big enough to be walking around on the tree branches near the nest. To Jon's de-light this one was also adopted by the foster parents. Later Jon observed all three fledg-lings flying near the nest. This is an impor-tant finding, as we have long believed that hawks would kill any offspring other than their own.

A majestic bird of prey, the Red-tailed Hawk is the most common and best known of Iowa hawks. The back and head are brown-ish, the tail red, the breast white, and the belly darker-streaked. Eighteen inches long with a 4½-foot wingspread, it is the largest and most powerful of the buteos, those large-bodied hawks with broad wings and short, broad tails. It is practically the only hawk still nesting in the Red Rock area along the Des Moines River.

Once the Red-shouldered Hawk and the Sparrow Hawk were common breeding birds here, but no more. They, and in fact all the hawk species except the Redtail, are either on the Audubon Society's blue early-warning list or on the red endangered-species list. Most feed at the top of the food chain, either on aquatic or avian species, thereby picking up dangerous amounts of persistent pesticides. Redtails feed almost entirely on mice and other small mammals that are apparently less contaminated, although the Missouri Conservation Commission found their white-tailed deer carrying heavy dieldrin concentrations which lowered the survival rate of the fawns.

Using a 20-power scope, I search for Redtail nests the last of March or early April before the trees leaf out. For a number of years I located four pairs of nesting Redtails in this township, but in recent years I have not located more than one or two pairs. They are not maintaining their population here. Why? If they are still being illegally shot, it is certainly not by the farmers who have come to appreciate the value of the big birds. Or is it the poisons?

The Redtail is a good hunter. I have observed at least three methods of hunting. There is the lofty-soaring flight with keen eyes looking for prey far below. Another method I have watched quite often along railroads: the bird simply perches atop a telegraph pole and patiently watches and waits. Less often I have seen them sailing low over a meadow, much like the hunting method of the Marsh Hawk.

I am always surprised that this big hawk is so fond of small prey. One researcher examined 173 stomachs and found that 131 contained mice, often three to four in a stomach and five in one. The prey includes mammals, mainly rodents, birds, snakes, amphibians, insects, and dead animals if not too ripe. In the total ecology of an area, the Redtail definitely plays a beneficial role.

It is thought that the pair mates for life, but if one is killed the survivor mates again. They are devoted mates and devoted parents. The pair indulge in courtship flight, soaring circles high in the air and often a series of slanting up-and-down V-shaped dives. This continues more or less during the entire nesting season.

The female lays two to four white eggs splotched with brown, bigger than hen eggs, perhaps only two here as I have never seen more than two nestlings to a nest. The incubation time is about thirty days, and the nestling life is at least forty-five days. As fledglings they are fed and trained to hunt for at least one month. Altogether, the parent Redtails devote close to four months of each year to rearing their young.

During the past April, May, and June, I often walked the mile into the refuge to watch the Great Blue Herons at their big stick-nests. Always there was a Red-tailed Hawk nearby. I scanned all the trees with a 20-power scope but never located a hawk nest.

The herons were not endangered by the hawk. As it occasionally circled over the rookery, the herons reacted by standing up but did not leave their nests. They simply gang up on a predator and drive it away if it is foolish enough to approach an occupied nest. A vicious jab of the spear bill could kill or disable a Redtail. The heron nests were never left unattended. One goes out fishing then returns and relieves the bird on the nest. This continues throughout the twenty-eight-day incubation period.

On May 9, George Ceolla, *Des Moines Register and Tribune* photographer, and I walked into the rookery. We discussed the Redtail, that it was a yearling with a dark tail, that they do not breed till two or three years old. I saw two little gray heads in one big heron nest, and I watched the Redtail holding a snake in its talons for half an hour as if waiting to bring it to nestlings. Again on a later date in May, I watched the two little gray heads in the nest and always one Redtail off to the east.

One evening in early June, Ruth Brisendine, Knoxville Cub Scout den mother, and I walked in, settled down with 20-power scope and 7 × 50 binoculars. Suddenly two young birds stood up in the nest, and they were not young herons! At last I had found the Redtail nest that I had hunted in vain all spring. The yearling was the male parent—I found only one other record of a yearling parent in the literature.

And what beauties they were. One was slightly larger than the other and more sedate. The slender one put on an aerial act for us. At dusk as we left, the Red-tailed female returned to the nest, the first time I had seen her.

On June 9 I walked in again, wondering if the young hawks would still be there. They were and their behavior was much the same. I named the quieter one Placid—it was standing on some sort of prey, tearing off chunks of meat and eating it. The slender one was performing one aerial act after another, flapping furiously, rising about 2 feet, coming down, dancing out on first one branch then another. I realize these flapping exercises are purely utilitarian, preparing and strengthening the bird for flight, but I was fascinated, gazing in admiration at this graceful ballet dancer clad in white leotards doing a beautiful dance on a rustic stage high in the air. And I had a name, Pavlova.

On their last day in the nest, June 10, Herb

Dorow of Newton enjoyed their antics and shot several rolls of film. By June 11 both birds were out of the nest. Pavlova had disappeared and Placid, the slowpoke, had flown only 15 feet to another deserted heron nest.

Larry Stone

Rough-legged Hawk

The winter of 1974–75 was one of marked contrasts, an almost total absence of the tundra and boreal land birds and the winter finches—the crossbills, Pine Grosbeaks, redpolls, and Pine Siskins. I received only one report of eight redpolls at a Knoxville feeding station. Mary Felsing and I netted and banded the only Pine Siskin we saw all winter. Purple Finches were the one exception, reported from all over the state.

The major mouse-lemming predators, the Snowy Owl and the Rough-legged Hawk, irrupted in full force. The Snowies moved south through the midcontinent in a spectacular irruption, probably numbering in the low thousands and reaching as far south as Waco, Texas, and Opelika, Alabama, delighting residents and tourists there for more than a month.

Thanks to *Des Moines Register* readers, eighty Snowies were reported to me in Iowa, the report later printed in the Iowa Ornithologists' Union's *Iowa Bird Life* (June 1975) and the National Audubon Society's *American Birds* (June 1975).

In the East the expected invasion fizzled, with only a thin trickle as far south as Pea Island, North Carolina. The previous winter, 1973–74, was a major flight year in the western states, with only fair numbers in Washington and Oregon in 1974–75.

The Rough-legged Hawk flight saw record numbers from the Great Lakes, the Great Plains, the Midwest, and the northern Great Basin all the way to the Carolina barrier islands, the Florida Keys for a "first ever" record, the Gulf Coast, southern Texas, with the comment "too numerous to report in detail," Chihuahua, Mexico, and southwestern California.

This large buteo, noted for its broad, rounded wings, robust body, and broad fan tail, is about 19 inches long with a 4- to 4½-foot wingspread. The Roughleg is polymorphic; that is, it has two color phases, but it is not sexually dimorphic (both sexes are colored alike). Of the 287 species of diurnal (day) birds of prey worldwide, 53 are polymorphic, and they are mostly open-country hawks that hunt small ground-mammals. Biologist Dennis Paulson of the University of Washington theorizes that polymorphism is advantageous in the taking of prey.

The dark-color-phase birds are dark brown all over except for a long white streak in the flight feathers and a broad black band across the end of the white tail. The light-phase bird, the one most often seen here, has a dark back, creamy head and throat streaked with brown, and a broad brown or black band across the belly; the wings are creamy underneath with a conspicuous black wrist (bend) and black tips on the flight feathers and a broad black band on the end of the white tail. The tarsus (leg) is feathered.

The Roughlegs are the last of the hawks to migrate each fall, usually the last week of October through mid-November, pushed south by the first snows in the far north. In a recent letter, Dr. Pershing B. Hofslund of the University of Minnesota, Duluth, reported "a fall flight of 60,000 hawks of all species through the Duluth funnel, with a nice November Rough-leg migration."

While touring the Red Rock area, I watched a Roughleg perched in typical very erect posture in a treetop overlooking an old field of tall grass, the brown-streaked creamy head and breast and broad, dark brown belly band glowing in the bright sunshine. These hawks of the far north are less wary than other hawks, and this one was undisturbed by our presence not too far away.

Soon it began a low, leisurely, graceful flight across the grass, similar to the quartering of a Marsh Hawk, revealing the creamy white undersides of the wings with black wrists and the white tail bordered with the broad band of black. Once it must have seen a meadow mouse, and there it literally hung suspended in the air, hovering, with wings vibrating rapidly, but the mouse evidently disappeared into a den, for the Roughleg did not drop its long feathered legs to pounce but instead continued its slow flight around the field and returned to the perch tree.

On December 2, I watched a light-phase Roughleg perched in a treetop on the Coal Creek bottom, occasionally flying low over a grassy field, returning each time to one of the treetop perches. Soon I observed a dark-phase bird almost completely black, soaring quite high. The light-phase bird took to the air, soaring too but keeping well below the dark bird that soon turned and disappeared to the southeast. The light bird resumed its alternate perching and flying low over the field. Was this a peaceful defense of a winter feeding territory or just play? I don't know.

Dr. Milton Weller a few years ago found them using communal roosts, three to eight birds, in big cottonwood trees north of Ames. His pellet analysis revealed the typical small mammal prey items, mostly mice, including two unusual items, the fur and tooth of an ermine (white weasel, *Mustela erminea*), rare in northern Iowa, and a bog lemming (*Synaptomys cooperi*), rare in that area, too.

Gary Schnell, in his comprehensive De Kalb County, Illinois, study, found six to fifteen birds in each of six communal roosts; four were in Norway spruce and two in big deciduous trees, all in farmyards except one in a cemetery near homes. More nocturnal than other hawks, Roughlegs come in to roost in midwinter about five to fifteen minutes after sunset and depart mornings from seven to thirty-two minutes before sunrise.

These magnificent mouse-lemming predators are found nesting across northern Europe, Asia, Siberia, Alaska, and Canada. The nest of sticks often is built on rock cliffs or preferably in tall trees.

The large sets of four, five, or rarely six pale greenish white eggs are variously spotted or splotched with dark brown ecru and sometimes a rich purple and pale lavender. Incubation is twenty-eight days. The young are covered with a thick white down, buffy on the head and back. During the ten-week nest life

they are fed a rodent diet, rarely a few birds, by the devoted parents.

In Labrador, the Roughleg is called Squalling Hawk because of the noisy protest made when the nesting territory is invaded. However, the courtship song is described as a soft, musical, plaintive whistle, very pleasant to hear.

Carl Kurtz

American Kestrel

Bats in the belfry? No, it's a pair of kestrels (Sparrow Hawks) that nested in the belfry of the Methodist Church in Knoxville this past summer.

What a joy it was to have a pair of these small, graceful falcons nesting in the church tower. I was first informed of this highly unusual nest site by a Knoxville teacher, Sharon Greenlee.

During incubation, the Robin-sized, gray-winged male would fly in with a mouse for his brown-winged mate, calling "killy, killy" as he approached the nest. Otherwise, the pair would probably have gone undetected, as they are very secretive.

The usual nest site is a cavity in a big tree. My brother-in-law was pleased to have a pair nest for several years in a box elder tree near his double crib where they preyed heavily on mice, rats, and House Sparrows.

Occasionally, they nest in cubbyholes in old houses. One pair nested in a stovepipe in a deserted farmhouse in central Washington County. Sometimes in the West they use an old magpie nest.

In 1913, Althea Sherman did an intensive study of a pair in an old blind originally built for watching marsh birds on the Garnavillo Prairie near McGregor. The first egg was laid on April 27, and each alternate day another egg was laid until the sixth and last egg on May 8. The usual clutch is four to five. The eggs are beautiful—the ground color is white to pinkish white and nearly covered with minute brownish dots and a few larger spots or splotches.

In Sherman's study, four of the eggs hatched on June 4, 5 and 6. From the first, the hatchlings made a faint cry in response to the parent bringing food, also faint peeps like a baby chick. On June 13 they first showed fear, flattening themselves on the floor of the box when Sherman looked in.

At two weeks they could run quite well when placed on the floor of the blind, and a few days later they braced their backs against the wall and faced their foe. Soon Sherman was met with savage claws when she extended a hand to pick one up.

At sixteen days the difference in behavior of the males and females was evident. The males were more timid, but the females would spring forward, every feather on their heads standing out at right angles, mouths open and squawking, ready to claw and bite.

The parents dressed the meat and tore it into bite-sized pieces until the young birds were twenty days old. Thereafter, the hawklets tore up the prey and fed themselves. The nest life is at least thirty days.

On hatching, the hawklets are scantily covered with white down, which soon changes to heavy yellowish white down. The characteristic male and female plumage is evident in the juvenile plumage.

Without doubt, this little falcon, members of whose family are of worldwide distribution in the temperate and tropical zones, is one of the most charming of all birds. It is very playful. One winter day, while I was doing a winter bird population study in a cemetery, I noticed a Red-tailed Hawk perched on a fencepost near a pond in a nearby field. Hovering over it and occasionally darting down to touch the Redtail's back was a female Sparrow Hawk. This play went on for minutes with no response from the Redtail until, finally tiring of the game, the little falcon flew to the timber.

They are skillful hunters of small rodents and often are observed during migration perched on powerline poles along gravel roads and interstate highways. Their attacks on mice are as fast as the speed of light. Although the fur is quite short on the various species of field mice, the Sparrow Hawks spend considerable time plucking bits of fur before skillfully dismembering it, sometimes discarding the intestines. Birds are also taken, although the preferred food is mice and, later in the summer, big grasshoppers.

One day each spring Gene and Marilyn Burns of Jamaica come for a day of kestrel banding in the Red Rock area. Using a Bal Chatri hardware cloth trap covered with slip nooses of fine fish line and containing a live house mouse, we cruise along the roads until we see a kestrel perched on a power pole. Driving slowly past, Marilyn gently drops the trap on the shoulder of the road; then we drive on a short distance and park.

Soon the bird flies down and steps on the trap, and as it lifts to take off, the noose tightens on its legs and it is caught. We immediately back up and remove it from the trap. It is weighed, measured, banded, and then released. The band number is sent to the U.S. Fish and Wildlife Service, which may help us understand the movements and status of the species.

Larry Stone

Ring-necked Pheasant

In 1786 LaFayette sent several species of pheasants to George Washington for release at Mount Vernon. They did not become established, and it was nearly one hundred years later, 1881, before the Ring-necks brought from China were introduced successfully in the Willamette Valley in Oregon.

In 1887 a number of Ring-necks were brought from England and successfully introduced in New Jersey and in 1890 in Massachusetts.

The Ring-neck is a blend of cold-weather races of Mongolia and China where it has played hide-and-seek for centuries in the scant cover along the edges of the paddy fields. Well adapted to civilization, it has done best in the grain belt across the continent. It never has been successful in the red-earth states from Maryland southward.

In Iowa it first was stocked quite accidentally in 1901, when a storm blew down the fence of a private game farm near Cedar Falls, releasing 1,000 to 2,000 birds. It long has been suspected that greedy hunters assisted the wind that night.

In 1913 the Iowa Conservation Commission (ICC) established a game farm and gave Ring-neck eggs and chicks to northwest Iowa farmers to be reared and later released. They flourished and by 1928 were common across the northern counties.

Stocking was extended to southern Iowa, and the first pheasants I recall were south of Pleasantville in 1931. Slow to establish here, they did not become plentiful until the 1950s.

The first pheasant season, three half-days in October 1925, covered thirteen north-central counties. Through the 1940s and 1950s the seasons were twenty to twenty-five days long. Since 1962, the seasons have been longer, nearly two months, except in a restricted area in southeast Iowa.

Unfortunately, the ICC did not shorten the 1975 hunting season in northwest Iowa where the pheasants were nearly all killed in the January blizzard. Izaak Walton League members and farmers protested in vain.

Among a long list of very costly failures,

the Ring-neck is undoubtedly the most successfully introduced game bird in the United States. Roger Tory Peterson, in *Birds Over America* (1950), stated that the pheasant was at the top of the game bird list with 16 million taken each year, exceeding the yearly bag of all wild ducks, and that the amount of pheasant meat reaching the table each season equaled the meat of 50,000 beef cattle. He also predicted a population crash for South Dakota, where 6.5 million pheasants then were harvested annually, and he was right. Iowa now produces more pheasants than South Dakota, but Iowa too is beginning to have a population crash now that government-diverted and set-aside acres have gone into row crops.

In 1973, the American Rifle Association reported 10.5 million pheasants killed by hunters in the United States—down 5.5 million since 1950, reflecting a continent-wide decline.

According to Dr. Allen Farris, ICC research biologist and pheasant authority, "The quality and quantity of proper habitat determine the long-term trend of the pheasant population." Winter blizzards and rainy nesting seasons are minor factors. In 1954 in Winnebago County, 57 percent of the land was in nesting cover—hay, oats, pasture, and set-aside acres. Today only 10 percent is in nesting cover; 90 percent is in row crops which produces few, if any, pheasants.

Roadside cover is becoming more important as nesting cover, and herbicide spraying ("brownouts") of roadsides should be stopped along with early mowing of roadsides. Thousands of songbirds, Bobwhites, and pheasants use roadsides as nesting cover.

It is a pleasure each spring and far into the summer to hear the exotic cock pheasant crowing, flapping his wings in exultation as he collects his harem of demure hens. What a beauty he is, running across the road or flying down a hillside.

The courtship display is a strutting performance, showing off the gaudy plumage, puffing out the feathers, and seeming to inflate the scarlet air sacs around the eyes. "Tidbitting," the presenting of food to the hen, is but one of the tricks this busy polygamist uses to keep his harem near him and away from competitors.

The pheasant nest is one of the most difficult to find, the hen blends in so perfectly. A few years ago on August 12, the children and I were crisscrossing a red-clover field, netting and tagging monarch butterflies. We had covered the area several times when a hen pheasant exploded in flight at my feet. The nest of ten pretty olive eggs was the latest I ever found. If the first nest is broken up, the hen will renest, and many will renest a second time.

On August 2, 1975, Iowa State University student Ronald Heinen and his wife brought me a young pheasant found dazed and injured in the middle of a four-lane highway near Des Moines. The juvenile plumage of streaked buff and brown resembled that of the adult hen, so I named it Pretty Girl. Very shy at first, it would hide in the covered coop at the end of the 2-by-2-by-5-foot wire cage. Recovery was rapid and so was the growth. I did my own feeding experiment, offering a wide variety of foods. Of the green foods, grass and clover were eaten, but the favorite was the weed lamb's-quarter. Elderberries, pokeberries, pears, plums, wild and Concord grapes, red honeysuckle berries, the ball of purple berries of the *Smilax* (carrion flower) vine, the white berries of our native dogwood all were eaten but an "apple a day" was the favorite fruit. Cracked corn, seeds, and wheat also were relished.

Arriving home after dark one night, I put food in the cage for the morning and was surprised by a peculiar, hissing sound made by the little bird, almost like a snake. Perhaps a protective mechanism?

One day Carl Priebe, refuge biologist, stopped by and commented about "the pretty cock pheasant." He had noticed the few bright feathers just coming in on the neck that I had overlooked. Quickly I changed the name to Pretty Boy.

By mid-September he was becoming aggressive, making rushes at my hands as I put in food. In late October, with my apple tree nearly bare, I decided Pretty Boy with his expensive "apple a day" appetite should be freed on the Davis's game farm. There he could get a handout if necessary until he adjusted to living in the wild again.

Larry Stone

Ruffed Grouse

This charming story came from a reader, Mrs. Ray Hebrank of Dundee. She writes, "We live west of Backbone State Park, and have about 60 acres of woods for the birds to enjoy with us. We have a Ruffed Grouse which is very friendly. He has lived here at least two years. He started to come close when husband Roy was cutting dead trees for wood to burn.

"He seemed to be attracted by the noise of the tractor and chain saw. We can catch him, but we let him loose after a short while after taking his picture. He usually comes to meet us when we go to the woods. The Conservation Commission released grouse and wild turkeys in Backbone State Park, so he may be one that was hand-reared."

I checked with Commission wildlife biologist Donald Cummings at state headquarters, and he assured me that they had not released any hand-raised grouse there.

Next I checked Bent's *Life Histories of North American Gallinaceous Birds* and learned that early hunters called the Ruffed Grouse the "Fool Hen." I quote, "Ruffed Grouse that have never heard the roar of a gun and have not learned to know their dangerous human enemies, are often absurdly, almost stupidly, tame; whereas the birds that have been persistently hunted have developed such a degree of wariness and strategy as to make it difficult to outwit them."

Bent also included many accounts of grouse that became attached to humans. In one, the farmer would call, "chickee, chickee," and the hen grouse would come out of the woods and sit upon his knee, then fly to his shoulder and back to the ground, chucking contentedly all the while.

In another account, a hen grouse flew into the barnyard one fall, possibly to escape an enemy, and lived there about a year, mingling with the chickens but roosting by herself in an open shed. She became tame enough to eat out of the hand and even laid a set of eggs in the orchard, which of course did not hatch.

Yet another account tells of Billy, the cock grouse that challenged a tractor on the road. When the driver stopped, Billy pecked the man's trousers and permitted himself to be picked up and handled, perching on the driver's finger, wrist, or shoulder.

Ruffed Grouse are birds of the forest, and in New England they were called the Forest Pheasant. Nonmigratory, their range is given as the United States, Canada, and Alaska north to the limit of trees and south as far as Brasstown Bald in Georgia and Roan Mountain in North Carolina.

In Iowa, this beautiful little game bird thrives only in the northeast, although a few have been transplanted in Shimek State Forest in recent years. Their population has increased to the point that hunting is permitted each fall, but only in the northeast, not in Shimek.

The spectacular courtship performance, the drumming of the cock, has been known for at least 250 years, and published records date back to 1755 (George Edwards, *Philosophical Transactions of the Royal Society of London*). The cock erects the cinnamon-red striped tail like a turkey cock and raises the ruff of feathers around the neck while standing on a fallen log. He then struts about and drums by clapping the wings against his sides faster and faster, which sounds like thunder at a distance. Some reports say the drumming can be heard at half a mile.

The female lays her eggs in a depression on the ground, lined with leaves and pine needles. It is usually at the base of a tree or stump, under logs, bushes, or brush piles. The nine to twelve pinkish buff eggs are incubated twenty-three to twenty-four days by the female.

The chicks are able to run about soon after hatching. They have reddish brown down on the head and back with yellow underparts and a black line running from the eye to the nape.

Their preferred food is aspen, but they eat a wide variety of foods which include nuts and seeds, buds, blossoms and leaves, and at least fifty different fruits. Over 80 percent of their diet is vegetable, with insects taken during the summer.

Hunting seasons are carefully controlled. Hunting is not the cause of today's population declines. The cutting of forests is a factor, and these birds are subject to many diseases. The late Dr. A. O. Gross did a comprehensive study of 2,000 birds from Quebec to Virginia and west to Minnesota. He found that they are subject to at least six infectious diseases and about twenty-five parasites.

The northeast Iowa birds have been studied in recent years by Conservation Commission biologists and currently comprise a healthy population. Where farmers raise few chickens, blackhead and coccidiosis practically disappear from wild turkey and grouse.

Over rugged terrain, the hunter finds this fast-flying bird a difficult target and is lucky to bag one bird for every three shells fired. However, my neighbor who has hunted these birds in Manitoba says they are much easier to hit than Bobwhites.

Larry Stone

Greater Prairie Chicken

One Christmas the National Audubon Society encouraged an "Antique Christmas Bird Count Experiment," in which birders covered precisely the same territory covered in an early count of fifty years ago. Since the Red Rock Christmas Bird Count included within the 15-mile diameter circle the farm 5 miles straight east of Pleasantville where I lived during early childhood, I was inspired to recall the Christmas of 1913, when I was five years old. The many trees in the 2-acre windbreak grove are all gone except one lone cottonwood. The house and all buildings have been leveled except the old barn, once red with white stripes, now all white.

You may ask how a five-year-old could possibly remember the birds of that long-ago winter. Simply repetition—each day Mother remarked about the birds seen. That winter the big thrill was the flock of a dozen prairie chickens wintering in Dee Galvin's field of corn shocks across the road north. Each day Mother would comment, "There are the prairie chickens" or "I haven't seen the prairie chickens all day. They must be in the north end of the field today." During a snowstorm she would remark, "The prairie chickens are safe inside a corn shock today with plenty of corn to eat."

The birds were a beautiful sight on the snow-covered field. The males were as big as chickens, 14 or 15 inches long, brown-and-buff stripes around the entire body, with long tufts of feathers on each side of the neck and orange air sacs on the sides of the head, inflated only during the courtship booming ritual. The slightly smaller female lacked the fancy headdress of the male. That was the last flock of prairie chickens I ever saw. Mother reported the last lone bird she ever saw on the Myers farm 2 miles south of Pleasantville in 1929. They were doomed with the turning of the prairie sod. Today, prairie chickens are being reintroduced with the hope they will once again occupy the Iowa landscape.

Carl Kurtz

Killdeer

The Killdeer is the most widely distributed and best known of all our shorebirds. Unlike most of the group, it is not confined to the borders of lakes or along the coasts but is found in pastures and dry uplands, often far from water.

It is the first of the shorebirds to arrive in spring and the last to depart in fall. Here I have known it to arrive in late February and depart in late November. They migrate by day, also by night.

Its scientific name, *Charadrius vociferus*, was well chosen. It is indeed vociferous, a very noisy bird, but the "killdee, killdee" call is a welcome one, often given as it flies over a field.

The pair bond is established on the southern winter grounds, and for the first four to six weeks after their arrival they spend a large part of their time in courtship. John James Audubon said of the courtship ritual, "It skims low over the ground, or plays at a great height in the air, particularly during the love season, when you may see these birds performing all sorts of evolutions on the wing."

The courtship flight is spectacular to watch. As they feed, one bird suddenly springs into the air, followed by the mate, and flies higher and higher in the wide sweeping spirals, all the time uttering its "killdee" notes until the two disappear from sight. Finally the calls cease and they literally plummet straight down to earth, alighting where they had taken off.

The Killdeer nests in the open on the ground, often in a pasture. No fancy nest, just a few straws or small stones in a hollow. However, many early farmers will recall flushing one from a nest at the base of young corn while cultivating with a team of horses. I once had that experience and quickly pushed the cultivator shovels wide to avoid covering the eggs with dirt. A couple of years ago wildlife photographer Herb Dorow sent me a picture of a Killdeer nest and eggs in the gravel of a tarred roof in downtown Newton.

The number of eggs is usually four, buff-colored and splotched with brown, quite pointed on one end. The eggs are quite large as these are precocial birds; that is, the young are able to run about as soon as the down dries. The incubation period is twenty-four to twenty-eight days, and both sexes incubate. If you walk across a pasture with baby Killdeers nearby, you will be given the "wounded bird" act to draw attention away from the babies, the parent dragging one wing and giving a wild alarm note.

With pastures fast disappearing, Killdeers are not as numerous as they were during my childhood. Their economic value is great as 98 percent of their food is insects. In a western alfalfa field, 383 alfalfa weevils were found in one Killdeer stomach. The Killdeer is a friend to humans, feeding almost entirely on insects, beetles, grasshoppers, bugs, fleas, worms, snails, and other invertebrates such as centipedes, ticks, and crustaceans.

They winter in the southern states, usually leaving Iowa in October. Once in early November I counted twenty-five in a clover field on the Iddlings farm south of Pleasantville.

Linda Kurtz

Upland Sandpiper

Although it was formerly named Upland Plover, this gentle, now rare bird is not a true plover. It belongs to that group of sandpipers that breed and feed in prairies and grasslands high above sea level and is now known officially as the Upland Sandpiper.

The Olaus Muries found them nesting in Alaska 200 miles north of the Arctic Circle, and they are found in suitable habitats across Canada and in Nova Scotia and south in the United States, through the central states.

They, too, are long-distance travelers, making a 14,000-to-15,000-mile round-trip to winter in the Argentine pampas, returning in spring to nest in North America.

In pioneer days they numbered in the mil-

lions before the market gunners, after exterminating the Passenger Pigeons, turned to the slaughter of the Upland Sandpipers. Long ago in Argentina they were widely distributed over 50,000 square miles of coarse native grassland. Laws now protect them in Canada and the United States but not in South America, where they are still hunted. Contributing to the decline in the United States is the loss of habitat and their hunting insects, not only in clover fields and pastures but between rows of corn and soybeans that have been treated with deadly insecticides.

For years they returned in early April to nest on three farms in the Pleasantville area and along the south border of Red Rock Refuge, but not one has been seen for the past five years. In late May, Herb Dorow, Newton wildlife photographer, called to tell me he had seen them in Roberts Creek Park just north of Lake Red Rock, and he had proof, a photograph. There is sufficient grassland in the park and an adjacent farm to meet their needs. May their tribe increase!

Look for them in spring perched atop a fencepost with spread wings slightly uptilted, sometimes giving the "qua-a-ily" call. They are best known for their prolonged, rather sad whistle. Fred J. Pierce of Winthrop, for thirty years the editor of *Iowa Bird Life*, described the courtship flight song beautifully.

"On still wings, these large birds circle slowly about, usually so high as to be mere specks in the sky, and give their shrill penetrating whistle which will carry nearly a mile, depending upon the wind and the altitude of the whistler. First there are a few notes sounding like water gurgling from a large bottle, then comes the loud 'whip-whee-ee-you'—long, drawn out, and weirdly thrilling.

"On more than one occasion I have seen

the bird, after circling at such a height as to be almost out of sight, close its wings and shoot to earth like a falling stone."

The incubation time has been given as seventeen days, but I never found a nest before the clutch of three or four spotted eggs was complete. For five summers beginning in 1919 I watched the pretty mottled black, brown, and buffy downy young on a farm west of Drakesville in Davis County. Precocial birds, they are able to run about and feed almost as soon as hatched. My father identified them and told me he had once seen large flocks of them before the turn of the century. A few years ago a pair had young in the Mockingbird field near Pleasantville, and I was soundly scolded for over an hour by the parent bird flying overhead calling "quip-ip-ip-ip" over and over again, so distressed that I finally left the field.

Only once, in 1959, have I had the rare privilege of watching the flock gather and feed prior to the long migration. In a cattle pasture of red clover they began to congregate in early August, with eight on August 13, and by August 18 they numbered nineteen. All had departed by August 29 except two that lingered on until September 5, an unusually late date according to Woodward H. Brown's "Annotated Checklist of the Birds of Iowa."

I have never banded an Upland. The only ones banded in Iowa in recent years were the five banded by the Homer Rineharts near Marshalltown in 1964.

There are, of course, many wild superstitions about birds. Ornithologist George Lowery reports an old French idea in Louisiana that those who ate the delicate flesh of the Papabotte, the French name for the Upland, were imbued with extraordinary sexual prowess.

This superstition may have accounted for the slaughter of tens of thousands there prior to the 1918 passage of the Migratory Bird Treaty Act, which gave protection to all shorebirds except the snipe and woodcock.

Larry Stone

American Woodcock

On May 10, 1972, Wesley Jones, park conservation officer in charge of Elk Rock State Park—located on both sides of Lake Red Rock just east of Mile-long Bridge—phoned inviting me to view a rarity, an American Woodcock nest in his park.

Driving in I was thinking that all my life I'd loved these bluffs and now they're part of a great state park, so very beautiful this spring for everyone to enjoy.

Wesley led the way to the spot on the bluff overlooking the lake. There in the bluegrass and buckbrush beneath a tall tree, the little woodcock sat on her nest. She made not a single movement as we walked closer and closer.

At 3 feet we stopped to admire her, the big black eye set well back on the head, the cross bar markings on top of her head, the very long bill, and the "dead-leaf" brown-and-tan protective coloration. Finally I stepped to 2 feet and then she flushed.

The four beautiful eggs, buff-colored with brown spots, are quite large, true of all precocial birds—their young are able to run about almost as soon as hatched. (Altricial birds on hatching are almost naked and require feeding in the nest for one to three weeks.) The incubation period of twenty-one days is long compared to the ten to fourteen days of the altricial songbirds.

Never before had I seen an American Woodcock nest, and it was a thrilling experience. There are records of nests in city parks but they are rare—this shy, secretive bird had selected a highly vulnerable site within 100 feet of a picnic shelter. Usually they nest in remote wooded areas.

Probably the site was chosen because of good feeding grounds nearby. Nocturnal feeders, they probe with the long beak for angleworms, their favorite food, in damp, boggy ground near streams or springs. The long bill is well supplied with sensitive nerves with a highly developed sense of touch.

It can detect the movements of a worm in the soil and capture it by probing. During dry spells, it turns over leaves searching for grubs, insects, and larvae. It is a voracious eater and has been known to eat more than its own weight in twenty-four hours.

The courtship evensong flight of this ridiculously dumpy little 8½-inch-long bird with the long bill, the short legs, and almost no neck is truly spectacular.

For a very short time during the breeding season the male forgets his retiring ways and comes out into the open to put on a show soon after sunset, as twilight approaches. On moonlit nights it sometimes continues through much of the night.

Sometimes, but not always, before the courtship flight the male struts around on the ground with short tail erect and spread, with bill pointing downward and resting on his chest, uttering now and then a rasping note "zeeip" or "peent," according to this passage from Bent's *Life Histories of North America Shorebirds*: "Suddenly he rises and flies off at a rising angle, circling higher and higher, in increasing spirals, until he looks like a mere speck in the sky, mounting to a height of 200 to 300 feet; during the upward flight the wings whistle continuously rather like 'twitter, itter, itter, itter' repeated without a break. Then comes his true love song—a loud three-syllable note—sounding like 'chickaree, chickaree, chickaree' uttered three times with only a short interval between the outbursts; then the bird flutters downward, zigzagging and finally volplaning down to the ground near his starting point. He soon begins again the 'zeeip' notes and the whole act is repeated again and again."

Native only in the eastern half of the United States, the American Woodcock in Iowa is found mostly in the eastern half of the state. It is a member of the snipe family of North American shorebirds.

John Beamer, refuge manager, has at least one known nesting area on the south side of Red Rock Refuge. Jack Coffey, refuge manager in the Chariton area including the new Rathbun Refuge, has them nesting each year there. Dale Stufflebeam, fisheries biologist, told me of watching a mother woodcock with four miniature young bobbing along behind her as they crossed a road in Lake Wapello State Park several years ago, but he never found a nest in his twenty-one years there.

The woodcock is an aristocrat among game birds. Some hunters call it the Timber-noodle.

Its flesh is delicious, a gourmet dish. Unfortunately, restrictions were not placed on hunting, especially by the market gunners, soon enough. Today there is no open season in Iowa and shouldn't be in any other state. Excessive hunting of a bird too easily hunted, summer shooting in the North, and wholesale slaughter during a long winter season in the South very nearly exterminated this bird in the late nineteenth and early twentieth centuries.

Carl Kurtz

Ivory Gull

Gulls

The term sea gull is definitely a misnomer if there ever was one. There are far more gulls inland in North America than along the sea coasts.

The Laughing Gull, our most numerous one along the Atlantic Ocean, has been recorded rarely over Iowa, first at the Engineer Cantonment north of Omaha in 1823, at Coralville Reservoir in 1977, and near Red Rock Dam in September 1989, and on Red Rock Lake in July 1991.

A European species, the Lesser Black-backed Gull was seen at Red Rock Lake in December 1984 and December 1987, also October and November 1989. Very rare, this brought out-of-state birders, some by plane. Franklin's Gull, seen at Red Rock, has limited fall migrations with fewer in spring, with their pretty pink breasts. However, the first week of September 1970 an estimated 100,000 to 500,000 hawked insects, following gleaners from Pella to Carlisle, a spectacular sight.

The lovely little Bonaparte's Gull is fairly rare in summer or winter. It is the only gull nesting in trees, sometimes seen in Churchill, Manitoba, with the nest always placed on the leeward side of an evergreen.

The only record of the Mew Gull in Iowa is at Lock and Dam 14 on the Mississippi River.

The Ring-billed Gull is our most common one at Red Rock, and in December 1982, we still had at least 4,000 and Saylorville had 2,000. With freeze-up, they disappeared for a few weeks and are largely absent in summer.

The Herring Gull is more common along the Mississippi River in winter as it is at Red Rock, with late dates there of May 22, 1973. They are almost always seen on Christmas Bird Counts in December, especially along the Mississippi River.

Thayer's Gull achieved species status by the American Ornithologists' Union in 1973, when it was separated from the Herring Gull complex.

The Iceland Gull, all white, was first seen by Pete Petersen at Lock and Dam 14 in December 1974 and January 1975. Our only one at Red Rock was identified by Ann Johnson and Tim Schantz on March 13, 1991. There were three on the Mississippi River near Davenport in early January 1992.

The Great Black-backed Gull, a very large four-year gull, is a huge bird, 30 inches long with a wingspread of 5½ feet and a massive bill. It has been seen several times, mostly in late fall and spring at Red Rock from 1978 through the 1980s. It is expanding its breeding range from Greenland and Labrador south to North Carolina and winters casually to the Gulf Coast.

The Black-legged Kittiwake is rated as an accidental, with the first specimen taken in Des Moines in November 1931 (DuMont). In recent years it has been an annual winter visitor at Red Rock, Coralville, Saylorville, and along the Mississippi River. It was first seen at Red Rock, November 23, 1988, and has been seen each year thereafter including January 1992 (Tim Schantz). A highly pelagic bird, it prefers to winter at sea along both the Atlantic and Pacific coasts.

Sabine's Gull is an accidental, with two records at Saylorville and two early records, 1891 and 1894, near Burlington. None have been seen at Red Rock. A beautiful gull, Beth Brown and I enjoyed one at Saylorville.

The Ivory Gull is our rarest at Red Rock, with one seen and photographed by Tim Schantz below Red Rock Dam, December 23, 1990, to January 2, 1991. It attracted many out-of-state visitors. Tim's photo of the bird riding a small ice floe was spectacular.

The Glaucous Gull is a huge predatory gull, a little smaller than the Great Black-backed, with a heavy body. It nests in the far north and is a five-year gull with no black plumage. The winter adult has a pale gray mantle. It reached Red Rock nearly every

winter. The first winter bird is snow-white, a very beautiful bird.

A. Cruickshank/Vireo

Franklin's Gull

In the fall of 1969 small Franklin's Gulls appeared by the thousands. They followed the big combines in the farm fields and at night roosted on great floating rafts on the waters of Lake Red Rock.

"One afternoon I watched a revolving, soaring column of Franklin's Gulls arrive and land on the wheat, surely a thousand. They spread out the full length of the field and began jumping 2 and 3 feet into the air. It was a curious phenomenon, and it took me several minutes to realize that they were catching grasshoppers. They swept the entire wheat field, then all rose gracefully and flew upstream to the next wheat field and repeated the performance.

"So we have a modern-day version of the Mormon story, but with one difference. Our gulls came from a greater distance than the California Gulls, saviors of the Mormons. Theirs came not from the Pacific, as commonly believed, but from the nesting islands in the Great Salt Lake, whereas ours came from hundreds of miles north."

The Franklin's Gull migration route is generally west of us, with fewer seen during spring than fall. This is the only gull that breeds in the northern hemisphere and then migrates south of the equator to spend the winter along the west coast of South America from Peru and Chile to Patagonia, feasting in the rich waters of the cold Humboldt current.

These small gulls, less than a foot long, have a color change in the fall and again in the spring. During the fall they wear a gray cap on top of the head, dark slate-gray upper wing surfaces and back, white neck, breast, belly, and tail, and dark red bill, legs, and feet.

In the spring, the change is dramatic. The entire head is black with white eye-rings, and the breast and belly are a lovely rosy pink.

Late one March, Beth Brown and Betty Eis of Osceola and I toured the full length of the refuge and lake, ending at the bay just south of the Corps of Engineers' headquarters at the south end of Red Rock Dam. It had been a most successful birding day, and so at this last stop we rather nonchalantly looked over the flock of Ring-billed Gulls loafing on a patch of rotten ice and checked the diving ducks in the water.

Suddenly we were no longer tired, no longer nonchalant. We had discovered three pretty little black-headed gulls with rosy breasts. The sighting was a first for Beth and Betty and equally thrilling for me, because I often miss the Franklin's Gulls during the spring migration. With the white-breasted Ringbills for contrast, there could be no doubt about the identification. They were Franklin's in nuptial plumage.

The next evening we attended the Rolling Hills Audubon Society meeting in Indianola and, of course, boasted about all the birds at Red Rock, but we considered our three Franklin's Gulls our best find.

Naturally, I had to share all this bird life, so I phoned Gene and Eloise Armstrong and they came from Booneville the next day. We had another good day of birding but found only one of the Franklin's mixed in with Ringbills at the west base of the dam. The other two were probably on a fishing foray.

Actually, these little gulls prefer insect food to fish. Our pioneer ancestors, watching these birds gleaning insects behind their plows, called them Rosy Gulls or, more often, Prairie Doves.

Most of them nest in Canada now, but in 1899 Dr. Thomas Roberts, author of the two-volume *Birds of Minnesota*, described the nesting habits of a huge colony at Heron Lake in southern Minnesota. The floating nests covered an area three-quarters of a mile long and 300 to 400 yards wide. The two or three eggs hatched in about twenty-one days.

Dr. Roberts described the chicks as pink-footed, pale-billed little balls of down that had a tendency to wander from the nest and drift far out in the lake. Then the parents and their neighbors raised a great hue and cry, attempting to get the chicks to return. Failing this, a parent grabbed each chick's neck in its strong beak and tossed the chick back 3 or 4 feet toward the nest. After several such tossings the chick often was bloody and exhausted but soon recovered.

There are no orphans, because other parent birds readily adopt neighboring chicks, often with protests from real parents. Sometimes one pair will acquire ten to twelve chicks, which they feed and treat with tender, loving care.

Linda Kurtz

Ring-billed Gull

Each fall hundreds of Ring-billed Gulls can be observed loafing on old Highway 14 east of the Mile-long Bridge and cruising over the waters of Lake Red Rock searching for small fish. This highly gregarious behavior occurs during spring and fall migration. From southern wintering grounds they migrate to breeding grounds in the northern United States and Canada.

Their nests, placed on the higher elevations of island shores, are made of dried grasses and weeds, lined with finer grasses, and often decorated with feathers. The two to three eggs are a pretty pinkish buff or greenish buff with splotches or spots of brown. After a twenty-one-day incubation period, the mottled buffy or gray chicks remain in the nest a few days but are fed by the parent birds until able to fly and fish for themselves.

In pioneer days they were often observed following the plow, where they picked up worms, grubs, grasshoppers, and occasionally field mice. Here at Red Rock they feed only on small fish.

Usually silent, the voice is not musical. When alarmed they utter a shrill piercing note, "kree, kree," like the cry of a hawk, but when fishing the cry is an occasional soft mellow "rowk."

Carl Kurtz

Black Tern

Beginning in August, that greatest of all flyways—the Mississippi flyway—comes alive with migrating birds. Peaking in September and October, it finally ends with a thin trickle at Christmas.

At its height, the migration is like a winged flood pouring down the vast Mississippi Valley and its tributaries. It is in this flyway that a greater number of species can be seen than anywhere else in the world.

So it was with high hopes on August 14, 1974, that Dr. Birkenholz, a native of Prairie City, now an ecology professor at the University of Illinois, and I toured the Red Rock area. We began with the Runnells mud flats and potholes at the west end of the 25,000-acre

Red Rock Refuge and ended our foray with the tailwaters below the huge Red Rock Dam.

Shorebirds (plovers and sandpipers), the earliest of all migrants, were everywhere on the mud flats, fifteen species. Dowitchers, the most amusing, probing for insects with their long bills, resembled the rhythmic stitching of a sewing machine. A Willet flew over, displaying beautiful underplumage.

Among hundreds of birds feeding on the Pinchey mud flats, we watched three rare Stilt Sandpipers and a few Semipalmated Plovers wearing one black band across the white breast instead of the Killdeer's two black bands.

Great Blue Herons were numerous, but only a few Great Egrets, Mallards, Blue-winged Teals, Wood Ducks in eclipse plumage, and three Green-winged Teals were there. One Least Tern was the rarest species seen. A few Turkey Vultures floated effortlessly overhead.

We enjoyed a simple picnic lunch on the very top of Red Rock Bluff with only the Indigo Bunting singing his "sweet, sweet, chew, chew" song. Bird song was no longer a necessity now that the nesting season had ended (except for the Goldfinch which never builds its first nest until near the first of August).

Below the dam was a beehive of activity, both human and avian. Happy people, fishing, camping, sightseeing and happy birds feeding and twittering. About two hundred Cliff Swallows were feeding young in the mud-jug nests plastered on the concrete face of the great dam. I counted 135 nests.

At least five hundred Black Terns were swooping with infinite grace back and forth over the water, their black bills pointed downward to feed on the insects and occasionally pick up small fish floating on the water. Others were resting on the sandbar in the middle of the rushing water.

Some were still in summer plumage, with black head and body and charcoal gray wings; some were already in winter plumage with a dark gray patch on the head, gray wings, back, and tail, and white underparts; others were moulting with a variety of spots and splotches.

Gulls and terns belong to the family Laridae, but the gulls are heavier-bodied with wider wings and square tails. The terns are slender, gull-like birds with narrower wings and forked tails, and when they search for food, the slender, sharply pointed bills are pointed downward.

Of the sixteen North American terns, only five reach Iowa. The Black Tern (found also in Europe and parts of Asia) is one of the smallest, only 9 inches long with a 35-inch wingspread; the largest is the 20-inch-long Caspian Tern, white with a black cap and large bright red bill with a 53-inch wingspread (usually seen during September).

In this area we see Black Terns only during migration. Some springs, but not every spring, they followed my father's plow, swooping down to pick up insects and grubs. Then Father would tell us he had seen the "seagulls" and predict a dry year. His is the only such prediction I ever heard. Actually, the Black Terns reach Iowa every spring.

The Black and the Forster's terns both nest in marshy areas of northern Iowa. Dr. Milton Weller, formerly of Iowa State University and now at the University of Minnesota, did a comprehensive study of their nesting habits at Rush Lake near Ayrshire and Dan Green Slough near Ruthven.

He found that these two terns did not compete for nest sites. The Forster's Terns used higher and drier sites while the Blacks preferred lower, wetter sites such as broken-down muskrat lodges.

Nor do they compete for food. The Blacks eat insects, only rarely picking up a small floating fish. They feed on mayflies, dragonflies, and other aquatic insects, whereas the Forster's eat fish as their main food with only a few aquatic insects.

Nesting begins the last week of May. The shallow cup of a nest is constructed of a few dead reeds or rushes. From one to four brown spotted eggs are laid, and incubation time is twenty to twenty-two days. Precocial birds, they are able to run about upon hatching but usually remain at the nest site several days.

Carl Kurtz

Yellow-billed Cuckoo

Almost everyone has heard the call of the rain crow, "kow-kow-kaw-kow-kawk," but fewer have seen this shy, secretive bird.

Old-timers believed that this bird's call meant rain and often it did precede a rainstorm. However, it sometimes sings at night and while in graceful flight. One elderly lady argued with me, insisting that it was the call of the Mourning Dove. She had heard it for seventy summers yet had never seen one.

This slender, elegant bird is the Yellow-billed Cuckoo, 11 inches long with olive-brown upper parts and a touch of rufous on the wings. The throat, breast, and belly are white, but the underside of the tail is black with three rows of conspicuous white spots, the best means of identification at a distance. The name is derived from the lower yellow mandible; the upper mandible is black and curved.

Found throughout the United States and southern Canada in summer, it arrives here about May 5, but it has been seen the last week of April. The latest fall date is October 21. It winters in tropical Central and South America.

The Old World cuckoos are parasitic, never building a nest, always laying their eggs in other birds' nests. However, the two in Iowa, the Yellow-billed and the Black-billed, are not brood parasites although they occasionally lay an egg in the nest of another species. I once found the egg of a Yellow-billed Cuckoo in a Red-winged Blackbird nest.

Bent's *Life Histories* reports eggs found in the nests of the Robin, Catbird, Dickcissel, Wood Thrush, Cedar Waxwing, and Cardinal. To quote Roger Tory Peterson, "It is among some of these occasional or partial parasites that we find hints about the intriguing questions of how the parasitic habit evolved."

Although listed as a common breeding bird in Iowa, the cuckoo population has declined in recent years, and it has been placed on the Audubon Society's Blue List in California, indicating a serious population decline there. I do not find as many nests as ten to fifteen years ago, but John Beamer reported them numerous in Red Rock Refuge the summer of 1972 after the nesting season had ended.

The flimsy platform nest of twigs, often so thin the eggs can be seen through the bottom, is located from 2 to 15 feet up, rarely as high as 30 feet. The usual number of pale greenish blue eggs is three or four, seldom five, with an incubation time of fourteen days. The newly hatched baby birds are homely little black creatures but amazingly strong. It is difficult to remove them from the nest, their feet seem locked to the nest twigs.

I remember what a shock it was the first time I looked in the gaping mouths of begging cuckoo nestlings. The mouth linings were black, and I was positive they were dying.

I had never heard or read of a black mouth-lining. In fact, there is little information in ornithological literature on mouth color. (As a taxonomic character, the color of mouth lining of tiny nestlings is of some importance. I have looked into a bluebird nest box and seen three bright yellow gaping mouths and a red one. I know full well that the red one was a cowbird, for all thrush babies have yellow mouth-linings.)

The appearance of cuckoo nestlings gets worse, not better, as they sprout quills eventually 2 inches long before the horny sheaths finally burst all at once in a single day, and the perfect feathers unfurl on about the eighth day. They look just like their parents—brown above with creamy underparts and ridiculous stubby tails about one-quarter inch long. I visited a nest of little "porcupines" once and returned in a couple of days to find that, like flowers, they had bloomed into perfect little cuckoos complete with bad tempers—they hissed at me, an odd buzzing hiss!

Among our most beneficial birds, they have voracious appetites with a decided preference for caterpillars. Hairy, spiny, or poisonous,

it doesn't matter in the least to the cuckoo, it eats them all. Often the stomach will have a pincushion lining of stiff caterpillar hairs, and sometimes an intestine will be pierced by one of the spines. The poisonous spiny caterpillars of the Io moth, a silk moth rare in Iowa but common in the southern states, are eaten in great numbers with no ill effects.

One observer watched a cuckoo eat forty-one Gypsy moth caterpillars in fifteen minutes and another consumed forty-seven tent caterpillars in six minutes. The cuckoos are the only birds that will eat the wooly bear caterpillar, larva of the pretty little apricot-colored isia isabella moth.

They are quite fond of many wild fruits such as cherries, grapes, blackberries, raspberries, and elderberries. I have captured several in mist nets for banding as they flew in to eat ripe mulberries.

Carl Kurtz

Screech Owl

Three successive winters a little gray-phase Screech Owl spent nearly every day perched on a south branch of the big cedar tree in my neighbor's yard. Undisturbed by traffic, it sat like a sphinx, perhaps secure in the knowledge that it was perfectly camouflaged.

This is the owl of my very early childhood, for in the big grove of soft maples, box elders, and cottonwoods surrounding our farmhouse there were plenty of cavities suitable for the nesting of this little 8-inch-long bird. Its funny little wailing cry was the first owl "song" I ever heard.

It is a shame that the name "screech" was attached to this little eared owl, for screech it does not. The love song heard during March is a quavering wail or whinny—a tremulous whistle that slides down the scale so that it

has something of a sad quality. Thoreau at Walden described it as "a solemn graveyard ditty, the mutual consolations of suicide lovers remembering. 'Oh-o-o-o- that I had never been bor-r-r-n' sighs one on this side of the pond. Then 'That I never had been bor-r-r-n' echoes another from the far side of the pond."

Unlike our large owls, the Great Horned and the Barred, the Screech has adapted well to humans, nesting near homes in town or country in any suitable tree cavity and occasionally accepting a nest box. Over the past sixteen years a pair has alternated nesting one year in a hole in the old silver maple at the second house east and the next year in the old maple half a block west of my home. Occasionally they missed a year.

The three to seven white eggs are incubated twenty-six days, and the nest life is about thirty-two days. Both parents feed the young, but the male supplies most of the food during the first ten days while the female covers the young.

On hatching, they are most appealing, wearing a coat of snow-white down. In the deep South they are called Shivering Owls, perhaps because they shiver from cold until they are in their thick gray woolly down after about two weeks.

One summer the fledglings from the west maple were perched for a few days in an apple tree overhanging the sidewalk. I was dive-bombed and bill-snapped by the "flying tiger" parents every evening that I walked that way. They never hit my head but came within 18 inches.

For many years on the farm, my parents had them nesting in an old box elder beside the driveway. This was a particularly vicious pair, dive-bombing within a couple of inches of any human head. They stayed around for weeks, not about to leave such a rich food supply of mice and sparrows. I recall my father's remark that the owls and snakes were worth their weight in gold to the farmers.

The preferred food is mice, rats, and other rodents, also a long list of songbirds—they are not desirable in a bird sanctuary—frogs, toads, crayfish, fish, snails, lizards, snakes, and larger insects such as the cecropia moth. A unique feature of owl digestion is the formation of pellets which are not passed through the intestinal tract. Pellets are regurgitated and contain the indigestible portions of their prey—bones, skull, fur, and feathers.

A study of the composition of owl pellets is very rewarding, revealing much about the ecology. Debbie Baldcock, a Pleasantville student, recently did such a study of Long-eared Owl pellets collected under evergreens in old cemeteries in Red Rock Refuge.

Dr. John Bowles, professor of biology, Central College, Pella, identified the animal species in one series of pellets and suggested that Debbie's findings should be submitted for publication in *Iowa Bird Life* or the *Iowa Journal of Science* since no such study of Longears had been done in recent years in Iowa. The rarest species was a southern bog lemming.

I examined a series of the Screech Owl pellets collected in each of the three winters. With three old barns and several open garages along the street, I was not surprised to find that 96 percent of the food was House Sparrows plus a few juncos and only 4 percent mice.

Enemies include the larger owls. Cats and other predatory mammals sometimes kill a Screech Owl on the ground while it is taking a mouse or other prey. Squirrels often destroy nests. Sometimes the owls are killed by cars along a highway.

The modern and most efficient method of doing an owl count is to play a tape recording of owl calls at night, stopping every quarter-mile. Most owls will respond (but not the Long-eared) to a hoot of their species. Jim Rod and other Ames Audubon Society members obtained a record count of twenty-seven Screech Owls in Ames on one Christmas Bird Count.

Larry Stone

Great Horned Owl

All breeding birds follow a set ritual of pairing, establishing a territory, courtship, nest building, egg laying, incubation, nestling care, and finally teaching the fledglings to hunt food. If one link of the breeding cycle is broken, usually the entire ritual is repeated.

Our earliest nesters, the Great Horned Owls, also our dominant bird of prey, carry on a noisy, hooting courtship during the month of January. The usual nesting territory includes an old Red-tailed Hawk nest 25 to 50 feet up in a big tree in a remote area. This owl never builds its own nest, it simply usurps an old one.

From early February 1977, I observed the pair that usurped last year's Redtail nest located 25 feet up in a wild cherry tree southeast of Pleasantville. Incubation began with the first of the two round white eggs, otherwise the first egg would freeze. The second egg was laid two or three days later. All through the incubation period, which lasts twenty-eight to thirty days, the slightly smaller male faithfully hunted rodents and fed the female each night.

On hatching, the two owlets, about baby chick size, were covered with snow-white down from the tiny bluish hooked beak to the talons on each toe. They made faint peeping sounds on the first day, but the eyes did not open for at least a week. The older owlet was noticeably larger, having a two-day head start.

Now the father was really overworked, with four mouths to feed, while the mother covered the tiny owlets. For the first feeding the mother tore off tiny bits of mice or rabbit, including fur and bones, feeding them gently. The owlets ejected small pellets containing fur and bones through the mouth.

When sufficient rabbit fur accumulated in the nest, the two owlets kept warm by huddling under the fur, releasing the mother to go on short hunting forays. At two weeks they were about one-third grown and covered with thick, buff-colored, woolly down, darker on the back. At three weeks they were half-grown, with the primary wing feathers beginning to burst their sheaths. During this period they have voracious appetites. The appetite slackened when the wing feathers were fully grown.

By mid-April they were full-grown except for the primaries, secondaries, and tail. As I watched they walked clumsily around the old nest, preened vigorously to remove feather sheaths, and did wing-flapping exercises. In another week they were out of the nest and on the ground, being fed by the parents. (They are incapable of sustained flight until eight or nine weeks old.) This is normal for Horned Owls and in remote areas they are safe. Do not rescue young Horned Owls unless in an area overrun by cats, dogs, and people. Notify your local conservation officer, as it is illegal to have any wild bird in your possession unless you have both federal and state permits.

Although the Horned Owls prefer nest sites remote from humans, there are exceptions. A pair nested in a haymow northeast of Pleasantville in 1972 and again in 1973. Julie Worthington, daughter of the farmowner, made careful observations from the laying of the first of the two eggs.

On March 17, the first egg hatched (the second egg never hatched). The tiny white owlet died when less than one day old. During that day Newton wildlife photographer Herb Dorow, using hay bales as a blind, photographed the mother owl and baby.

Chemical analysis done at Iowa State University revealed death was due to persistent pesticide residues. DDT and dieldrin have since been banned because of their long-lasting and devastating effects on wildlife.

This spring one lady reported Horned Owls rearing two young in her haymow. In 1975 the owls had first nested there.

However, the most unusual nest site of this spring was the one in a residential area of Prairie City located 20 feet up in a big hole in a silver maple tree on busy, noisy Highway 163 at Ruby Conley's home. Several people in the neighborhood, including Louis Birkenholz, Lulu Bos, and Ruby, had been aware of the hooting of these big owls beginning in January, but they were totally unaware of the nesting until mid-April.

On April 21, Lulu Bos, a former Pleasantville schoolmate, phoned to tell me of the nesting and Louis Birkenholz notified me by letter. I immediately phoned Herb, asking him to photograph the full-grown young owls in their nest hole. This proved to be a well-nigh impossible feat due to the height of the nest, the poor light, and the buff color of the young owls blending perfectly with the wood background.

On April 23, Maxine Crane of Pleasantville and I spent a pleasant afternoon observing the young owls high above the roaring traffic while visiting with Ruby and Lulu. "Snappy," the older one, glared angrily at us with great golden eyes and furious bill-snapping although we were fully 50 feet away, while the younger one bill-snapped only occasionally.

On April 27, Herb phoned to tell me that Snappy had left the nest April 25 and had successfully flown to a nearby tree after an initial flight to the roof of the big house across the street, also that Snappy had been successfully photographed.

The current rage for owl figurines reinforces the fact that owls have held a fascination for people since prehistoric times. The first picture of a bird easily identified by species is that of a Snowy Owl scratched on the wall of a cave in France called Les Trois Frères. Cro-Magnons were aware of bird migration.

In folklore, the owl is universal—mysterious birds of the night. The "wise old owl" dates back to the Greeks who associated the owl with the Goddess of Wisdom, Athena, and an owl appears on one side of ancient Greek coins. In China it was the bird that snatches the soul, also the symbol of thunder and lightning. It is a canny creature in Eskimo lore and a worthy antagonist to the sly coyote in the lore of our southwestern American Indians. To the Mayans, the Screech Owl was the symbol of Ah Puch, the God of Death, who ruled the lowest level of the underworld. Owl mummies were found among the grave furniture and were common on tomb paintings in Egypt.

Carl Kurtz

Snowy Owl

My love affair with the Snowy Owl began long years ago when, at the age of ten, I saw my first one on our farm. It was many, many years before I saw another.

Armed with federal and state permits, I hoped for years that I would someday have the opportunity to care for one. My chance came on January 30, 1981, when John Klein, executive director of the Clarke County Conservation Board, phoned to ask if I could take a Snowy that had been picked up in a farm field near Osceola by two Marshalltown sportsmen, Roger Henze and Gary Lang.

The bird was taken to an Osceola veterinarian where it was examined, photographed, weighed, measured, and dusted for ectoparasites. The emaciated, starved bird weighed only 2 pounds and had a minor wing injury, but that was of no importance. It had not been shot.

The bird was kept overnight at the veterinary hospital, then Klein and Beth Brown delivered him to me the next morning. I named him Nicki, a corruption of the scientific name, *Nyctea scandiaca*, *nyctea* being Greek for nocturnal.

I placed him in an outdoor flight cage, but he would not eat. I quickly decided he was too weak to survive the cold weather, so I brought the bird into my seldom-used west living room, put him into a big pasteboard box, and hand-fed him a cup of meat. When I lifted the box top, he would look up, open his beak, and hiss at me, and then I would pop a piece of meat into his mouth.

Later that evening he upchucked all the food, undigested, and I knew I had a very sick bird on my hands. I phoned my veterinarian, Dr. Mark Poell, and asked for instructions. I told him the bird's mouth-lining was white instead of pink.

Poell advised me to give the bird a small piece of rabbit every two hours, and I did that through the night except for four hours' sleep. The bird retained the food, and the next morning Dr. Poell made a house call and told me the bird's body was so depleted that food alone would not save him. He prescribed high-potency vitamin and mineral tablets and an antibiotic to control intestinal infection.

So, around the clock I fed Nicki pieces of rabbit or pheasant and gave him the medication. I averaged about four hours' sleep each night, plus a few catnaps now and then.

When I walked into the living room at 3 A.M. three days later, there was Nicki, standing in the middle of the floor. He had broken the cord tie on the box and jumped out. He was obviously regaining his strength. He ate his meat out of a dish after that, although I still hand-fed him the medication.

The bird was incredibly beautiful, snow-white except for dark gray markings on wings, breast, and head. The legs and toes, top and bottom, were covered with furry feathers. Only the great golden eyes, beak, and talons were bare. The down on the breast was 2 inches thick.

The only sounds he ever made were hissing and bill-snapping, except for a couple of occasions when he let out a scream—once when Carl Kurtz was taking pictures and once when my cousin, Sandra Strong, was here with her camera.

George Ceolla, a photographer for the *Des Moines Register and Tribune*, came in February. He set up lights and shot a roll of film, but the flashbulbs and strong lights didn't disturb Nicki at all. He would stand flat-footed on the couch without moving, except to turn his head from side to side. The *Des Moines Tribune* published the picture on February 7 and the Associated Press picked it up.

The following Sunday, Ceolla brought his daughter and her friend to see Nicki. Ceolla wondered if Nicki had gained any weight, so I brought out my kitchen scale, padded the platform with terry cloth, and Nicki stood quietly while we rejoiced. He had gained half a pound in one week, during which he had eaten two rabbits and three mice.

Dr. Poell made a second house call on February 11 and pronounced Nicki much improved. But he said for me to continue the medication.

On Saturday, Phil Benson and a photographer from Des Moines television station KCCI came and shot forty minutes of videotape, knowing they could use only ninety seconds of it on the newscasts, but they hoped CBS would pick it up. The network did, and I had reports from friends who saw Nicki and me on TV in Arizona and California.

Soon Nicki became the most photographed bird in Iowa. On Sunday, February 15, he had twenty-nine visitors, most armed with cameras.

Nicki and I began visiting schools. Hundreds of children in Pleasantville, Knoxville, Hartford, Carlisle, and Newton fell in love with the big white owl. Usually, he was placid and well behaved, standing on a tabletop. In Newton, however, he flew the length of two classrooms, affording the little second-graders a beautiful view of the snow-white wing linings as he passed over their heads. He also visited many club meetings where I had slide programs scheduled.

Nicki and I attended the Rolling Hills Audubon Society meeting in Indianola on February 24 and I gave a slide program, though he was the star of the evening. A bird-bander, Gene Burns of Jamaica, placed a U.S. Fish and Wildlife aluminum lock-on band, No. 599-2091, on Nicki's leg.

Three days later, the temperature rose to 68 degrees F and I put Nicki in a partially enclosed outdoor flight cage. He never used the high roost, but he liked the 3-foot-tall pedestal perch and he stood on the earthen floor, too. After all, there are no high perches on the tundra.

In early April I began to wish Nicki could be released in the far north and facetiously wrote Elizabeth Clarkson Zwart, columnist for the *Des Moines Sunday Register* and the *Des Moines Tribune*. I explained that I was looking for a "well-heeled fisherman, preferably one with a Lear jet, who could give Nicki a lift to Canada."

My request was printed in her column on April 8, and she received a phone call the next day from the wife of a fisherman with a Lear jet. Nicki was offered a ride to Canada in mid-May.

Then came the hassle to get a customs permit and medical certificates. Dennis Crouch, the U.S. Fish and Wildlife agent in Des Moines, requested a customs permit from the regional office at Fort Snelling, Minnesota. We never did get it; they claimed they never got the request.

Dr. Poell produced the necessary medical certificates, and we decided that Nicki could be released at Baudette, Minnesota, and that he could fly across the border himself. Crouch made arrangements with David Carpenter, a Minnesota conservation officer who would accept and release Nicki.

An hour after we delivered Nicki to the airport in Des Moines on May 15, he landed safely at Baudette to a welcoming committee usually reserved for visiting heads of state— reporters, photographers, airport officials, Carpenter, Minnesota, conservation officer Greg Spaulding, and others. Nicki was again the star, and he got in a few whacks at the young officers, biting Spaulding's finger and digging his talons into Carpenter's arm.

Finally, he was released at Carpenter's home 8 miles east of Baudette along the Rainy River. Did he fly straight north into Canada? No, he landed in a nearby field, where he was harassed by a crow. It took thirty minutes for the officers to drive the crow away. Nicki hunted over the fields for three days before returning to his native Canada.

My thanks to one and all for the kindness and food supplied during the Snowy's sojourn home.

Larry Stone

Hawk Owl

The crown jewel in Iowa's list of rare birds this bitter-cold winter of 1981–82 is surely the Hawk Owl, a first sighting for the state. The bird was first seen a few days before Christmas perched in the top of a tall tree near the intersection of two busy highways near Waterloo by two birders who travel that way daily. They paid little attention, thinking it was just another Barred or Great Horned Owl. On the Waterloo Christmas Bird Count, when the birders of that party focused on the bird, they quickly realized they had struck gold—a Hawk Owl, and a first.

As soon as it was placed on the Iowa Bird Hotline, hundreds of birders traveled across the state—some from other states—to observe the rarity. A very obliging bird, it remained in that area into early February, evidently finding the hunting good.

Swainson and Richardson in 1831 gave us the earliest account of this circumpolar species. "This small owl, which inhabits the Arctic Circle in both continents, belongs to a natural group that have small heads, no ear tufts, and imperfect facial discs and considerable analogy in their habits to the diurnal birds of prey [hawks]. It is a common species throughout the fur countries from Hudson's Bay to the Pacific, and is more frequently killed than any other by the hunters, which may be partly attributed to its boldness and its habit of flying by day"—this from Bent's *Life Histories*.

Actually, it breeds north into Alaska, east to Labrador, and south to southern Quebec and is resident in the great northern forests of poplar, spruce, pine, birch, tamarack, willow, and alder, broken here and there by small prairies and muskegs, its favorite hunting grounds.

Our North American Hawk Owl is much darker than the European bird. It is a little smaller than a crow. The head is small, the back and wings are dark mottled brown, the breast and belly are finely cross-barred, and the eyes and beak are yellow. Its very long tail, giving it a falconlike appearance, is often flicked and cocked up. When hunting, it hovers above prey much like the Sparrow Hawk.

In fact, while Gene and Eloise Armstrong of Booneville watched the bird one day in January, it flew out, hovered above its prey, then pounced down into the grass and snow. There it remained hidden for several minutes before swooping up again into the perch tree. It had made a kill, probably a meadow mouse, and had eaten it on the ground. Through the scope, the Armstrongs plainly could see the blood streaks on the yellow hooked beak.

In my childhood my mother taught me the names of all the common birds, but I heard the words Hawk Owl only twice. Once, sixty-five years ago, in northern Wisconsin, a Ruffed Grouse, frightened by an owl, flew into our picture window and died of a broken neck. My father and the farmer across the road observed the daylight chase. The neighbor, a Bohemian, told my father that it was a Hawk Owl and that he had seen them during bitter-cold winters in Bohemia.

Then one late afternoon in the winter of 1922 when we lived in Davis County, my father heard the hens squawking in the henhouse. He grabbed his gun and shot an owl clutching a white pullet in its talons.

He said it was a Hawk Owl like the one in Wisconsin, which is, of course, doubtful, but it was not a Barred Owl. The eyes were yellow, and it had no feather horns. The very first owl I had ever touched, I marveled at the soft feathers and the brownish, speckled breast that reminded me of a barred rock hen, although of a different color.

Hawk Owls nest in old snags, in woodpecker holes, and in old hawk and crow nests. The three to seven white eggs the female lays are incubated twenty-five to thirty days, and the owlets remain in the nest twenty-three to twenty-seven days.

The food taken varies at different seasons but usually includes mice, lemmings, and ground squirrels during the Arctic summers. When snow is deep, the bird often follows flocks of ptarmigans and lives on them. It also takes ermine, an expensive meal.

The birds spend the winter well within their breeding range in northern Canada and Alaska or move a short distance south. Only rarely has there been a major invasion into the northern United States. Bent, in *Life Histories*, records a memorable invasion in 1884 and another during the winter of 1922–23.

Carl Kurtz

Short-eared Owl

At first glance the Short-eared Owl might be identified as a Barred Owl. Its short ears are so very short, they are almost invisible, and it has lovely golden eyes while the much larger Barred Owl has brown eyes. A little over a foot long with a 40-inch wingspread, the soft plumage is a warm brown and buffy gold, with lengthwise belly stripes, a distinctive dark patch at the bend of the lighter under-wing, and a large buffy patch on the darker upper wing.

The face is most appealing; the large facial disk is outlined by a thin black circle. Black feathers surround the golden eyes. White feathers above the eyes appear to be eyebrows and below the black hooked beak look like whiskers. Buffy gold fur boots cover the legs and even the toes, leaving only the "fishhook" talons bare. The "fur" is actually very fine feathers.

The Short-ear is a cosmopolitan species found on every continent except Australia. They're found all over North America, breeding as far north as Alaska and wintering all the way to the Gulf. During my childhood they still nested in Iowa but no more. Now an uncommon winter resident, the Short-ear was very nearly exterminated by ignorant and ruthless gunners before Iowa finally passed a long-overdue law protecting all hawks and owls, effective July 1, 1970.

This is our only diurnal owl, a day hunter, but it prefers hunting in the early evening and through the night. Most owls like deep woods, but the Short-ear hunts the open fields, sailing and flapping low, occasionally hovering over one spot before pouncing on its favorite prey, one of the numerous species of mice and moles. Sometimes it is mistaken for the Marsh Hawk as their hunting methods over open fields are similar.

This owl nests on the ground, lining the nest with a few fine grasses and weeds, usually in the open but sometimes partly hidden by a clump of tall weeds or grass. The courtship flight and song are unique, unknown among the other owl species. Flying at a considerable height with slow-flapping wings, sometimes soaring, the male does a short dive, bringing the wing tips together beneath his body with short clapping strokes eight to twelve times in a few seconds, meanwhile doing a series of fifteen to twenty toots.

Large families are the rule, five to seven eggs the usual clutch, with incubation of twenty-one days beginning with the first egg laid. Consequently, nestlings vary greatly in size. One nest contained nine nestlings varying in age from three to fourteen days. On hatching, the owlet is covered with buffy white down above and pure white below. They remain near the nest six weeks but are able to wander about the grass from two weeks on.

To lure predators away, the devoted parents stage "distraction displays" full of sound and fury. They groan, chuckle and squeal, flap, flutter and tumble, feigning injury, the "wounded bird" act, sometimes resorting to the wing-clapping act of the courtship flight.

The Short-ear is a friend of people. One study showed that 75 percent of its food consisted of mice and that it was more of an insect eater than any other owl. One owl's stomach contained 59 grasshoppers, many beetles, and cutworms. About 15 percent of the diet consists of small birds. The late Paul Errington, from a study of pellets (owls regurgitate the indigestible bones, skulls, and fur of the prey), found the remains of 68 meadow mice, 115 deer mice, a Snow Bunting, and a meadowlark.

In winter, occasionally one will be seen perched motionless on a fencepost intently watching for a mouse. Once in January 1944, on a 45-mile trip to Osceola, I counted more than a dozen perched on fenceposts or sailing low over the fields, the last time I have ever seen them in numbers, probably a peak meadow mouse year. I am lucky to see one each winter now.

Three owl species—the Great Horned, Barred, and Screech—are permanent residents in Iowa. Another three—the Long-eared, Short-eared, and Saw-whet—are northern species seen here only in winter or during migration. The Barn Owl is a very rare permanent resident, and the Snowy Owl of the far north is an extremely rare winter visitor.

Carl Kurtz

Northern Saw-whet Owl

The smallest owl that reaches Iowa is called the Northern Saw-whet Owl. It bears the scientific name *Aegolius acadicus* and once was known as the Acadian Owl, a much more pleasing name than Saw-whet.

Only 7 inches long, it is a rare Iowa winter resident most often observed between November and January, with extreme dates of October 5, 1959, in Davenport, October 5, 1985, at Lake Rathbun, and May 6, 1988, in Lansing. Not many are seen in southern Iowa, but Donald Gillaspey reported two in Decatur County in recent years, and Beth Brown reported one in 1981 and another in 1982 in a pine grove in Clarke County.

In the Des Moines area, one was located in a pine grove near Big Creek during the winter of 1981–82, and this winter of 1982–83 Dick Mooney saw two in the same grove. The Waterloo–Cedar Falls birders reported two there in November 1982. Another was seen near Iowa City in December 1982.

Statewide, only four were found on the Christmas Bird Counts in 1981, one each at Marble Rock, Clinton, Cedar Falls, and Iowa City. Only once—in 1970—have we had one on the Red Rock Christmas Bird Count.

Secretive and nocturnal, the Saw-whet almost always is found roosting in pine groves, usually in small trees. A rare exception was one found in a small deciduous tree at the Ruan Center in downtown Des Moines. The photograph was printed in the *Des Moines Tribune* in December 1981.

Covering a quarter-century of bird banding in Davenport, Pete Petersen has handled and banded more Saw-whet Owls than anyone else in the Midwest. Only once have I had the pleasure of handling and observing a Saw-whet, so named because its song sounds like the whetting or filing of a saw blade. Some twenty years ago I had a net just north of Mother's kitchen window.

One November morning, before daylight, Mother called me and said she was sure there was a bird in the net, maybe a Screech Owl. I rushed out and carefully disentangled the little bird. It was an owl, so tiny that I felt sure it was not a Screech as I could see no ear tufts.

It was a lovely little Saw-whet, the first we had ever seen. The tiny bird was quite tame and did not fight as we examined it.

Saw-whets are beautiful brown on the head, back, and wings, with some streaking on the breast and fine white feathers like fur covering the legs and toes to the small, sharp talons. The facial disks surrounding the golden eyes are white from the black beak up over the eyes. Streaks radiate out from the eyes.

According to Bent's *Life Histories*, the Saw-whet's range covers all of North America. It winters from southern Canada to the central United States.

The breeding range extends north to southeastern Alaska and east to Quebec. The Saw-whet nests in deserted woodpecker holes after a brief courtship period. During courtship, the male serenades his mate with a "whoop-whoop" or "kwook-kwook" at the rate of three whoops every two seconds. The singing is especially persistent under a full moon at midnight.

The five to seven pure white eggs are laid at intervals of one every three days, with an incubation period of twenty-six days starting with the first egg. Consequently, there may be newly hatched and nearly full-grown young in the same nest.

The young are fed by both parents and remain in the nest from twenty-seven to thirty-four days, fed mainly on mice and small mammals. The hunger call is a hissing "t-z-z-z-z-z-z."

Larry Stone

Common Nighthawk

The goatsuckers are so called because of a European superstition that these birds with their tremendous gaping mouths sucked the milk from goats at night. Only two members of the goatsucker family, the Common Nighthawk and the Whip-poor-will, are found in Iowa although Charles and Darleen Ayres had the very rare Chuck-will's-widow nesting near Ottumwa, documented on tape, photographed, and banded three years ago.

Worldwide there are ninety-two species. All are nocturnal birds possessing fluffy owl-like plumage, but the feet are weak; they walk with difficulty. They sit lengthwise on a tree limb, not crosswise as do other birds.

The Common Nighthawk is dead-leaf brown, mottled above and barred below, with a white bib beneath the chin and a distinctive large white spot on each of the long, hawk-like wings. The body is Robin-sized, 9 inches long, but the wingspread of 23 inches is much greater than that of the Robins!

Roger Tory Peterson calls them "flying insect traps, capturing their quarry in cavernous gaping mouths" in the early evening before dark and throughout the night. It is not a hawk nor is it related to the hawks.

These small birds eat enormous quantities of insects. Beal reported finding 500 mosquitoes in one stomach, 650 plant aphids in another, and 60 grasshoppers in yet another. One stomach contained 2,175 ants, and several contained 50 different species of insects.

A few pairs nest in Pleasantville and Red Rock Refuge. In town they nest on flat rooftops on the square, but in the refuge they still nest on flat, bare sandy ground, using no nest materials at all.

The only ground nest I ever found was in June 1925. I was cultivating corn with a horse-drawn cultivator when the female flushed, and just in the nick of time I saw the two eggs at the base of young corn plants and shoved the plow handles apart to avoid crushing the eggs.

The courtship display of the male is a most ardent ritual. In the evening or early morning he flies in wide soaring circles above the female at the chosen nest site, swooping down almost to the rooftop or ground, then sharply up again. It is in this sharp upturn that the well-known "boom" is produced by the rush of air through the wing primaries.

Alighting near the female, he rocks his body, fans out his tail, puffs out his throat displaying the large white patch, and makes croaking sounds. That and the "peent, peent" call he makes while in flight constitute his singing. A musician he is not.

The two gray eggs splotched with dark gray are incubated by the female for eighteen or nineteen days, often in boiling heat. In Maine, Alfred A. Gross recorded the rooftop temperature beside an incubating bird, panting incessantly, at 130 degrees F, although the temperature at street level was only 98 degrees in the shade.

The female covers the nestlings all day long to ensure their survival, and she never leaves the eggs during the day. Both parents feed the young on insects for twenty-five to thirty days.

It is strange that the nighthawk took so readily to the flat city roof which only came into general usage around the mid-nineteenth century. Unlike the Chimney Swifts whose nest sites in giant forest trees were destroyed by early settlers centuries ago, forcing them into chimneys, the number of gravel sites on the ground suitable for nests of the nighthawks has not diminished, yet they quickly accepted our structures.

Perhaps the gravel-tar rooftop is not the safest place to nest, but it is much safer than the traditional ground nest sites. True, many nestlings fall off the roof every summer, but in the ground nest they are beset by a legion of predators—raccoons, opossums, foxes, minks, weasels, skunks, cats, dogs, and at least a dozen species of snakes.

However, their worst enemy in past years was humans. In the southern states during the nineteenth century, they were shot by the thousands for food or target practice during fall migration. Through educational efforts, the National Association of Audubon Societies finally persuaded legislators to pass laws around 1903 protecting these wholly beneficial insectivorous birds.

Linda Kurtz

Ruby-throated Hummingbird

In Brazil the hummingbird is called *beija-flor*, which in Portuguese means "kiss-flower," a name well deserved by these jewels of the bird world. We have only 1 species, the Ruby-throated, in the eastern United States and 13 species in the western United States. It is difficult for us to believe that this is the second-largest bird family in the Western Hemisphere, with 320 species. Only the American flycatcher, with 367 species, is larger.

More than half the hummingbird species are found in the equatorial belt 10 degrees wide across South America. Others are extremely hardy, dwelling at altitudes in the high Andes of Ecuador, Colombia, and Peru, where the thin atmosphere drops to freezing soon after sunset. Some solve this by sleeping in caves, even building their nests there.

Bundles of energy with high metabolic rates and enormous appetites, often consuming twice their body weight in syrup daily, they have developed a mechanism called torpor for conserving energy on cool nights. The pulse rate drops to 36 per minute. When very active, a hummer's heart may beat 1,260 times per minute, and at rest the rate is 480 beats per minute. Respiration rate is also high, sometimes 273 per minute. Body temperature ranges from 102 to 108 degrees F.

Their diet consists of both nectar and tiny insects, with a greater intake of nectar. Gram for gram, the hummingbird has the greatest energy output of any known warm-blooded animal. The noted hummingbird authority, H. Crawford Greenewalt, pointed out that if a human's expenditure of energy per unit of weight were as great as a hummer's, a person would have to eat 370 pounds of boiled potatoes or 130 pounds of bread each day.

Despite living at such high intensity, hummingbirds have surprisingly long lives. According to banding records, some live seven to nine years in the wild. In the Bronx Zoo and also in A. Ruschi's South American aviary, they averaged ten years and one lived fourteen years.

Some migrate, but many do not. The Chilean Firecrown breeds as far south as the tip of South America, in cold and bleak Tierra del Fuego, migrating northward during the coldest months. In western North America the Rufous Hummingbird is the greatest traveler, breeding from southern Oregon and Idaho to southeastern Alaska and in winter migrating far south into Mexico, a 2,000-mile journey. Our Ruby-throat is also a long-distance migrant, breeding as far north as southern Canada and occasionally reaching western Panama in winter, and some cross the Gulf of Mexico, a 500-mile non-stop flight over water.

Many, but not all, are tiny. The smallest bird in the world is the Bee Hummingbird of Cuba, 2¼ inches long, and half of this length is bill and tail. The largest hummer is the Giant Hummingbird of the Andes, 8½ inches long, the size of a Starling.

Best known for their beautiful colors, the hummers' fiery glitter is mostly due to structure rather than pigment. Iridescent colors are common among them. By changing position, the direction of reflected light may give the effect of two completely different colors of the same plumage. Often it depends on where you stand when observing the hummer as well as the source of light.

Its flight is also unique in the bird world. With great ease, it moves forward or backward, up or down, to right or left as well as pivoting. It also can fly upside down. When suddenly facing danger it can turn a backward somersault, dart a short distance with its wings in reverse and feet upward, then roll over and continue in normal flight according to Alexander Skutch in his delightful book, *The Life of the Hummingbird.*

The male puts on a spectacular courtship display. Swooping in a great wide U, at the lowest point he is just above the female who pays no attention at all, continuing to move from flower to flower sipping nectar.

In a recent letter, Dr. C. J. Hemmers of Oelwein described perfectly a Ruby-throat's courtship display above the prostrate body of a female temporarily knocked out after hitting a picture window. The observation was made July 7 while he and wife Virginia were visiting friends in Elkader whose lot slopes down to the Turkey River (they have Summer Tanagers too!).

A devoted spouse the male hummer is not, taking off as soon as the eggs are laid, assuming no parental duties whatsoever. I can't say that I blame him, for despite the brilliant courtship displays, the female is completely indifferent except for three or four days. And apparently his departure causes her no great grief. She goes serenely on, building the artistic nest 1 inch high and 1 inch deep, lined with plant down and covered with tiny pieces of lichen.

Alone she incubates the two little eggs about fifteen days, and alone she cares for the pea-sized naked nestlings, inserting her long tongue into their throats and pumping them full of nectar and tiny insects.

Recently J. B. Owen, bird columnist of the Knoxville, Tennessee, *News-Sentinel*, sent me two of his stories in which he described the nest of a Ruby-throat built on the end of a shelf bracket only 4 inches from the spotless white wall of a friend's home.

The tiny nestlings expertly handled their own nest sanitation during the twenty-day nest life. To quote J. B.: "Soon after the young hatched, specks began appearing on the wall when the youngsters were hardly larger than honey bees. The tiny birds were seen to hoist their rear ends and let fly with as much neatness—in the nest—as an old-time tobacco chewer spitting his amber at a cuspidor across the room."

From mid-August through September the migrants are welcome visitors at syrup feeders. Not many are seen here during the breeding season, although Maxine Crane had a few here in town and my young cousins, Bill and Sandra Strong, also had a few at their country home.

All is not peaceful at the feeders, however. There's always one little bully driving others away from its feeding territory. A year ago Maxine Crane put up a second feeder on the other side of her house. That undoubtedly became the most frustrated hummer ever as it zipped back and forth over the top of the house vainly attempting to hog both feeders.

Remember to measure the sugar accurately for your syrup feeder. It should be one part sugar and four parts water. Otherwise they will develop cirrhosis of the liver.

Larry Stone

Rufous Hummingbird

The fall migration of 1981 brought a big surprise to Thom Roberts of rural Indianola. He phoned to report that he found a Rufous Hummingbird among his flowers on August 28.

There are no records of this tiny bird in Iowa prior to 1981. Woodward H. Brown, in his "Annotated List of Birds of Iowa," did not mention the Rufous.

Be sure to check the hummers at your syrup feeders and flowers to see if you can spot the Rufous. The male is rufous, or copper-colored, on back, tail, and sides and has a brilliant red gorget, or bib.

The female is green on the head, tail, and back and has a finely speckled white throat. She is rufous on her sides and at the base of the tail.

Of the 319 hummingbird species in the New World, only 14 reach the United States and only the Ruby-throat, a long-distance migrant, is common in the eastern United States. However, the Rufous is the greatest traveler, covering more than 2,000 miles from its winter grounds in Mexico to its farthest north breeding grounds in southeastern Alaska, a tremendous journey for a tiny bird weighing only 3 or 4 grams.

It is common in the Far West, making its northward migration west of the Rocky Mountains, but the fall migration takes it farther east, mainly to the crests of the Rockies and the Sierra Nevada, wherever it can find blooming flowers. The males migrate first, often as early as July 1, while the females and young appear in late July and into early September.

The male's courtship display is spectacular, looping back and forth in towering U-shaped dives, often as high as 100 feet, all the while flashing the gorgeous red gorget in the face of the female as he speeds toward her.

Bent, in *Life Histories*, records this species as occasionally forming small colonies during the nesting season, with as many as twenty nests built in a small area, a fir grove or lower in huckleberry, blackberry, or alder bushes and long trailing vines. One nest hung from the end of a tall fern.

The beautiful nest is made of cottony plant down, profusely covered with lichens and bright green moss bound together with spiderwebs. An average nest measures 1½ inches across and 1¼ inches deep, with an inner cup about ⅞ of an inch wide and ½ inch deep.

The two white pea-sized eggs are incubated twelve to fourteen days by the female. The care of the young is done entirely by her, the father disappearing as soon as the eggs are laid. Feeding is done by the mother inserting her beak down the baby's throat and pumping a mixture of nectar and tiny insects.

Nest cleaning is no chore for the mother, as the nestlings practically stand on their heads and, with their tail ends at the edge of the nest, shoot the excreta several inches away from the nest.

The Rufous is the most hot-tempered, the most pugnacious of all the hummers. When they are not fighting among themselves, they declare war on nearby birds, swallows, thrushes, and even chipmunks, causing them to squeal in terror as they flee.

Great singers they are not. A twittering note, "chewp, chewp," is heard from the male during courtship and sometimes a "tut-ut-ut-ut-turre," and their wings make a humming sound.

Recent records include July 10–11, 1986, and July 29–August 3, 1987, at St. Olaf and August 18, 1987, at Clear Lake.

Carl Kurtz

Red-headed Woodpecker

The winter of 1980 was not a winter for northern finches, but it certainly was the winter of the rollicking, frolicking, boisterous Red-headed Woodpecker. They were found in unusually high numbers in our 1979 Christmas Bird Counts which cover areas 15 miles in diameter. A total of 1,757 were counted in the state, the largest number ever reported. Red Rock had 49 and Jamaica had 70 Red-headed Woodpeckers. The prize, however, goes to the Warren County Conservation Board in Indianola. On December 30, they counted 28 on their 80-acre Hickory Hill farm and 40 on the 160-acre Otter Hill farm, totaling 68 on 240 acres; that's an average of 1 bird for every 2.3 acres.

Why does this migratory woodpecker not go south every winter? I prepared a graph showing the populations recorded on Iowa Christmas counts for most of the past twenty years and found great peaks and valleys. For instance, Red Rock had only 2 on the 1977 count following the drought and only 15 in 1978.

I consulted Woodward H. Brown, author of "An Annotated List of the Birds of Iowa." With well over half a century of birding experience, Brown was of the opinion that the success or failure of the acorn crop is the deciding factor. Many acorns means many Redheads, and a crop failure means few, if any, Redheads overwintering here.

The noisiest of all our woodpeckers, the Redheads make the woods ring with their "char, char" or "quar, quar" calls. Having attracted attention, they play hide-and-seek by moving sideways to the back of the tree, then peeping around to see if you are still there, as if to ask, what are your intentions? They actually seem to enjoy playing games with an observer.

Long before I was five years old I knew this bold and dashing bird. Each spring it announced its arrival by drumming on the metal peak of the roof of our farmhouse. Mother would say, "Run and look, the Red-headed Woodpecker has come back." We ran and were overjoyed.

From that time until fall, our windbreak

grove was alive with the noisy chatter of these handsome birds. We watched as a nest hole was excavated, followed by a two-week incubation and a three- to four-week nest life. They were busy, industrious parents, such fun to watch.

Then came the glorious day when the young Redheads emerged. But they didn't have red heads; they had black heads and dusky breasts, not at all like their handsome parents. Such noisy fledglings they were—chattering, chirping, and cackling until it reached a crescendo.

In that long-ago time every farmstead had at least one pair of Redheads, but not now. Long before the last tree in the old groves fell or was bulldozed, the alien Starlings had usurped the nest holes.

The aesthetic value of the woodpeckers is great, but their economic value is even greater. They have a widely varied diet, about 50 percent animal matter and 47 percent vegetable. The insects consist of ants, wasps, beetles, bugs, grasshoppers, crickets, moths, and caterpillars. About one-third of all their food is beetles and 11 percent ants. They feed on the ground, in trees, and also fly-catch insects in the air.

The vegetable food includes corn and all the wild fruits such as cherries, apples, and pears. Acorns and beechnuts are favorite winter foods.

The tongue of a woodpecker is very unlike the soft pink tongue of a songbird. It is round, firm, pointed, and covered with backward-pointing hooks for fishing grubs from holes. It also can be extended 1 to 2 inches to reach a grub. The musculature of the head is specially adapted to cushion the shock of heavy pounding and drumming.

This species is blue-listed, having suffered drastic population declines in recent years. Factors in the decline are loss of habitat, insecticides, herbicides, road kills, and Starlings.

Carl Kurtz

Red-bellied Woodpecker

For constant enjoyment at the winter feeding station or in the winter woods you just can't beat the woodpecker. Three species—the Downy, Hairy, and Red-bellied—are permanent residents; two, the Red-headed and the Flicker, are summer residents though a few overwinter; and one, the Yellow-bellied Sapsucker, is a spring and fall migrant, rarely overwintering. All six are found in Red Rock Refuge during the winter months. The Pileated Woodpecker is found in big timber along the Mississippi and other major river valleys.

The Red-bellied Woodpecker is a Carolinian species, a bird of the South. It was not here to greet our pioneer ancestors with the "chur, chur" call but followed soon after.

By 1907, Anderson listed it as common in southern Iowa and rare in the north, but in recent years it has extended its range to the northernmost parts of the state according to Woodward H. Brown's "An Annotated List of the Birds of Iowa." It is only in the wooded areas of Red Rock Refuge, Elk Rock State Park, and Lake Red Rock that they are abundant—forty-five on one Christmas Bird Count, which may be near the carrying capacity.

A walk in the winter woods of Red Rock Bluff is always a pleasure. There I can be sure of hearing the Redbelly's noisy, cheerful "chur, chur," outdone only by the far noisier Red-headed Woodpecker who forgot to go south for the winter.

One Saturday recently I set one of the wire hardware cloth bird traps to catch a Starling. Instead, a male Redbelly walked in, stepped on the trigger platform, and the door dropped behind him.

As I brought him into the porch to band, I called to Junior and Christine, ages seven and five, children of my next-door neighbors, Kenny and Bonnie Simbro, to come and watch me band the handsome bird.

As I placed the numbered aluminum band on his leg, we talked about his temper, which matched the bright scarlet cap. He struck repeatedly with his long, straight, chisel bill, but held correctly, he could not reach my hand. I explained that he has very strong head and neck muscles and a very thick skull to absorb the shock when he hammers.

Occasionally he opened his beak, exposing the round, pointed tongue covered with tiny backward-pointing hooks for fishing grubs from the holes.

To aid in climbing he has four toes on each foot, two pointed forward and two backward, with long toenails, and his stiff tail feathers act as a prop. Rough tree-bark had already worn off the red tips of the belly feathers except those near his legs.

The red does not cover the entire head, only the cap and nape. (The female has red only on the nape.) The sides of the head, breast, and belly are a warm, creamy gray. The back is black-and-white zebra-striped—a ladder-back effect. The tail is also black and white.

I told them that he roosts in a hole of the big maple tree at our neighbors', Hubert and Chima Isenberg. In the spring the Red-bellied pair spends a week excavating a nest hole in a big tree. They take turns incubating the five to seven white eggs fourteen days and later feeding the young on insects and grubs.

However, the adults like a variety of food, including fruit and grain. At my feeders they eat suet, cracked corn, and apples, and they carry off food to stash in hiding holes for bitter-cold days of famine.

Did the Redbelly desert my feeders? He sulked for five days, visiting only the feeders of my five neighbors, but returned on the sixth day. The day after a December snowstorm he fed at five different suet feeders and two cracked-corn feeders and sampled apples wired to the Anoka apple tree and on my east window shelf. Frankly, he is a glutton, fat and plump, the best-fed woodpecker in the country.

Carl Kurtz

Hairy Woodpecker

On December 10, 1976, I removed a banded female Hairy Woodpecker from the north trap, brought her onto the porch, recorded the band number, 502-60619, and released her, noting that she was plump, her plumage well groomed, and that she appeared to be in good physical condition. Then I began flipping through my annual banding schedule expecting to find the number in the 1976, 1975, or 1974 records, as that is about the average longevity. To my surprise I did not find it, so I continued the search through 1973, 1972, 1971, 1970, 1969, 1968, and finally there it was, banded on April 25, 1967, an adult captured in a mist net in my front yard. The bird was at least ten years old! Next I checked through John H. Kennard's "Longevity Records of North American Birds," published in the winter 1975 issue of the *Bird Band-*

ing Journal, wondering if mine might be a record. The longevity record was twelve years and ten months, a Hairy banded by R. M. McCullough in Stillwater, New Jersey, May 20, 1961, and trapped again and released alive April 26, 1974, within the same area.

For a wild bird to live so long is a miracle today. Since Woodward H. Brown, in "An Annotated List of the Birds of Iowa," rated the Hairy a common permanent resident in 1971, it has suffered a population decline and is now on the National Audubon Society's Blue List. There are so many modern-day hazards—picture windows, traffic, insecticides, and loss of habitat through the death of all our elms and the clearing of timber for more acres of row crops.

The Hairy is a "big brother" edition of the Downy, white below and black above, white spots on the wings, with a broad white stripe down the center of the back. The male has a red patch on the back of the neck. It is much larger than the Downy, 7½ to 8 inches long, much more active, has a longer and larger bill, and has pure white lateral tail feathers, whereas those feathers are barred with black on the Downy. To aid in climbing it has two toes forward and two behind with long curved claws, and the stiff tail acts as a prop.

The long beak of the Hairy is hard as a chisel and is driven by powerful muscles in the head and neck. The very thick skull absorbs the shock. I always enjoy showing students the peculiar long pointed tongue covered with backward-pointing "fishhook" barbs especially adapted for snaking grubs out of holes. Insects make up 72 to 88 percent of its diet, with the remainder being fruits, nuts, and acorns. In Georgia, the slightly smaller Southern Hairy was very fond of the small black berries on our sour gum tree. Here I have netted them coming in to feed on ripe mulberries.

It ranges across all of North America from Alaska to the mountains of northwestern Mexico, where it is called the Chihuahua Woodpecker, and from the Atlantic to the Pacific wherever there is timber.

The Hairies excavate their nursery with powerful blows of the chisel bill, and it often serves as sleeping quarters later in the year. Not all birds' songs are as vocal. The rhythmic tapping or drumming of the Hairy is its courtship song. The three to six snow-white eggs are laid in the unlined cavity usually on a pile of chips. The incubation period is about fourteen or fifteen days, and the young remain in the nest three to four weeks. The defended nest territory is very important in spacing birds to assure enough food for the growing nestlings.

This species is noted for the difference in appearance of males and females. The dominant male is slightly larger and has a bill about 11 percent longer than that of the female. The woodpeckers are a hardy family. The longevity record of the Red-bellied Woodpecker is twenty years and eight months, the Flicker twelve years and five months, the Red-headed Woodpecker eight years and eleven months, and the little Downy ten years and five months according to banding records filed with the Bird Banding Laboratory, U.S. Fish and Wildlife Service.

Carl Kurtz

Eastern Phoebe

Each spring I eagerly search for the little bridge bird, the Eastern Phoebe, a plain little dull-colored bird, sparrow-sized, darker on the head than on the back, white to gray on throat, breast, and belly, a black bill, and no wing bars. The downward tail which wags is a sure identification trait. (The similar Wood Pewee has white wing bars.)

Arriving here in mid-March, they leave in September with a late date of October 30. They winter in the southern states, and when I used to live in Georgia, I would begin the watch for them in late October or November, where they spent the winter fly-catching on our lot or in the adjacent woods. Generally they were silent throughout the winter.

Of the ten members of the flycatcher family found in Iowa either as migrants or breeding birds, two are kingbirds, the Eastern and the

Western; one, the Great Crested, is a tyrant flycatcher; two are phoebes, the Eastern and the Say's, very rare in northwest Iowa; two are flycatchers, the Wood Pewee and the Olive-sided; and three are of the small *Empidonax* genus, the Acadian, Yellow-bellied, and Willow. All are difficult to identify unless heard singing. The flycatchers generally have olive to gray plumage and broad, flat bills and catch insects on the wing with a loud snap of the bill.

In my area the Eastern Phoebe has suffered a drastic population decline in recent years. Bridges, square culverts, and old buildings where I regularly found them nesting for fifteen years no longer have phoebes. The nest, eggs, and incubating female in the big culvert at the corner of the Mockingbird field were destroyed following a heavy rain three years ago. And they are gone from the old log house and all the bridges I have checked nearby.

It was therefore a joy to hear one singing its "fee-bee, fee-bee" near a park footbridge and to see one carrying nest material on April 21 while I was birding with Maxine Crane. A week later three Rockwell couples, Mr. and Mrs. Leo Peters, Mr. and Mrs. Walter Lorenzen, and Mr. and Mrs. Ernest Hitzhusen, and I were touring the Red Rock area, and one of the men located the neat little mud-and-moss nest containing one pure white egg. It was a first phoebe for two of the party and the first active nest for all, adding pleasure to a delightful day. On April 30 a bus load of members of the Ding Darling Chapter of the Izaak Walton League observed the nest and five eggs.

The substantial nest requires some thirteen days of hard work to build, seldom less. The female incubates the four to six white eggs fifteen to sixteen days. Bent's *Life Histories* re-cords occasional nests of eggs marked sparingly with brown spots near the large ends, but all the nests I have studied contained pure white eggs. Both parents feed insects to the nestlings during the sixteen-day nest life. The young fledglings are much prettier than the adults, having olive-green backs, yellow breasts and bellies, and two distinct buffy wing bars. Usually two broods are reared each summer.

One year on entering an old abandoned farmhouse to check a phoebe nest, I was greeted by a horrendous exhibition of snarling, growling, spitting rage as three young raccoons climbed the plastered walls of the old house. I beat a hasty retreat, leaving both the raccoons and phoebes in peace.

Many nests were in big square culverts, which meant wearing rubber boots and wading through the mud at the end of the culvert. Sometimes the mud was too deep, and those young phoebes remained unbanded. Occasionally a pair of Barn Swallows nested in the same culvert with a pair of phoebes, usually one nest at each end with no conflict between the species.

In 1966 Frances Phillips Metcalf notified me of the presence of Cliff Swallows nesting under a new concrete bridge near her farm home. Along with the forty-four mud-jug nests of the Cliff Swallows, we found one pair of phoebes serenely feeding the young in their nest, undisturbed by the furious pace of the Cliff Swallows flitting in and out. In the fall we collected several of the mud-jug nests for Frances's science project and also collected the phoebe nest.

The mud-jug nests were swarming with swallow bugs, an avian bedbug (*Oeciacus vicarius*), and a few were found in the phoebe nest. These bedbugs prefer Cliff Swallows, and their presence in the nearby phoebe nest was probably accidental. However, various ectoparasites, lice, and mites are the greatest enemies of the phoebe nestlings, often causing death or premature leaving of the nest.

Another major enemy of the phoebe is the Brown-headed Cowbird, a brood parasite. Many times I have gnashed my teeth in rage as I watched the little phoebes feeding a cowbird fledgling, the only survivor of the nest.

A. Cruickshank/Vireo

Great Crested Flycatcher

Of the thirty-three members of the flycatcher family found in North America, only twelve reach Iowa and three of these are very rare. Flycatchers have broad, flat bills and snatch insects in the air with a loud snap. Most have gray or olive plumage.

The three most common ones here are the gray-and-white Eastern Phoebe, which plasters its mud-and-moss nest to the walls of cement culverts, bridges, or old buildings; the black-and-white Eastern Kingbird, undoubtedly the most quarrelsome; and the cavity-nesting Great Crested Flycatcher, 7 inches long with a large olive head and back, gray throat, yellow belly, brown wings with white wing bars, and a bright, rusty tail.

In early May I begin listening for the loud "wheep, wheep." Flycatchers become very attached to a home territory, so I check the same spots each year along the Red Rock bluffs. Always I find them at refuge headquarters, the Dyer Bluff, Red Rock Bluff, several in Elk Rock and Roberts Creek parks and the Corps of Engineers land, all nesting in tree cavities.

The Great Crested is very territorial and where territories overlap, fights are frequent and feathers fly. The courtship consists of a male chasing a female around and around trees. Often she darts into a nest hole, then peeks out at the male. After mating, the nest building may take two weeks, the time needed to fill an 18- to 30-inch cavity with dead leaves, roots, hemp, string, fur, and feathers.

There is a true song sung only in the very early morning from 4:30 to 5:30. I have never heard it, but it is reported to sound like "coodle, queedle, coodle queedle." Margaret Morse Nice says it is given at the rate of twenty-eight to thirty notes a minute, often lasting twenty-eight to thirty-five minutes.

At the Felsing home in the woods near Red Rock Dam, a pair has nested in a bluebird box for several years. Almost always this bird uses a cast snakeskin as part of the nest material. This year the snakeskin is prominently placed on the front edge of the nest.

Is snakeskin a "scarecrow" used to frighten

away predators? Some birders think so, but other shiny materials, such as onion skin, waxed paper, and cellophane, often are used. Other species occasionally use snakeskin, too. I have found it in four wren nests in Iowa and in one Blue Grosbeak nest in Georgia.

On June 1, at the invitation of Lynn Padellford, I set up nets at the Felsings and banded birds for the Sams (good Samaritans) Campers from nearby Hickory Ridge Campground. One net was set in front of the flycatcher box in hopes of capturing and banding the adults to determine if the same pair returns each year, but no luck, too much sunshine on the net for these keen-eyed birds. (Rose-breasted Grosbeaks coming in for sunflower seeds outnumbered all others that day!)

The four to five, rarely eight eggs are white to pinkish white, heavily marked with brown to maroon streaks and blotches, very handsome eggs. The female does most of the thirteen- to fifteen-day incubating but is not a close setter and often one or two eggs do not hatch.

The nestlings are quite noisy, making sounds like early spring peepers when the parent birds approach with food. Both parents feed the young on insects for twelve to eighteen days. While most of the food is flying insects, larvae comprise at least 20 percent of the food. The parent hovers above the leaf as it picks off a larva. Butterflies and moths, wings and all, are jammed down the nestling's throats—even the Blue Jay is more considerate, carefully removing insect wings. At the Felsings we watched the adult drop down to the ground at the base of a tree and pick up an insect. The adults relish many wild fruits, too.

The range of the Great Crested includes all of the eastern United States and southern Canada to the Great Plains. By mid-September all have migrated to Central America south to Colombia.

Carl Kurtz

Horned Lark

On the 20-mile trip from Pleasantville to Red Rock Dam, Paul and Mary Felsing and I observed at least a thousand Horned Larks feeding on the shoulders of the highway, the only bare ground in the snowy landscape. Strictly ground birds, they never perch in trees. They fold their wings tightly after each beat, resulting in a distinctive choppy, undulating flight.

Attractive brown birds larger than a sparrow, they have two small black feather horns on the head, a black collar below the yellow throat, and a black whiskerlike crescent curving below the eye.

The northern subspecies has a yellow eye stripe, whereas our prairie race has a white eye stripe. The thousand probably were of the northern race forced south by a January blizzard with temperatures around 13 degrees F below zero, but we were not foolish enough to stop and haggle over subspecies identification. Probably there were a few Lapland Longspurs and Snow Buntings mixed in the flocks, as Dean Mosman of Ankeny had reported huge mixed flocks there.

The Horned Lark is the only true lark in North America belonging to the Old World Alaudidae family. (The meadowlark is not a true lark—it belongs to the blackbird family.)

Circumpolar, the Horned Larks are divided into forty subspecies, fourteen in the Old World and twenty-six in the New World. They range from the Arctic coasts of both hemispheres to northern Africa and South America, always on barren or short grassland.

How do they survive such bitter cold nights? Biologist Charles H. Trost found that during cold nights they dig roost holes with their bills, usually in the shelter of grass clumps or rocks just big enough for the bird to sit in with the back level with the ground surface. This prevents considerable heat loss in the cold and windy areas most frequented by the larks.

The Horned Larks are the earliest nesting songbirds in Iowa, and the entire breeding cycle is of unusual interest. Why should such a small bird choose to establish a territory in February, go through the courtship ritual, build a nest, lay eggs, and begin incubation by March 27? The answer is not known, but it undoubtedly lies far back in time before the ice ages. G. Pickwell, in his in-depth study of the Horned Lark, found that nesting was triggered by two consecutive days having a mean temperature of 40 degrees F above zero.

Like the famed Skylark of England, this lark also sings on the wing. The courtship flight song often begins at 4:00 A.M. and may continue off and on through the day with a final burst of evensong lasting well after sunset.

The male's climb to flight song is at a 60-degree angle into a strong wind or in an irregular circle if the wind is light, reaching a height of 200 to 800 feet. During the singing the bird flies in curves or circles. One May evening as I stepped out of my sister's farm home, I heard the beautiful tinkling song high up, a "jingling metallic sound like distant sleigh bells." It was too dark to see the bird or to watch his descent.

The female digs the shallow nest cavity with her bill, loosening and flipping away the soil, sometimes kicking dirt from the hole with her feet. Digging takes one or two days. The nest, sheltered by a tuft of grass or rock on the windward side, is woven of fine grasses, grass heads, feathers, string, paper, and shreds of cornstalks.

One feature is the "paving" laid on the open side of the nest, probably to cover the fresh dirt thrown out. Clods, pieces of cornstalks and corncobs, and dry cow manure are used in this paving.

The three to five eggs, gray with brown spots, are incubated eleven days by the female. I once found eggs that were quite dark, alive with brown spots. Many early nests are lost to late blizzards, and many early nestlings starve when the supply of insects is cut by snows. Heavy rains and agricultural activities destroy many nests, so that nest success is as low as 23 percent fledged and seldom higher than 48 percent.

Swallows

Good sites for the attachment of mud-jug nests help assure successful nesting seasons for Cliff Swallows in the Red Rock area. In 1971 more than 100 nests were plastered on Painted Rock Cliff upstream from Mile-long Bridge on the north side of the Des Moines River. In 1974 I counted 135 nests attached to

Carl Kurtz

Tree Swallow

Northern Rough-winged Swallow

D. & M. Zimmerman/Vireo

Fred Kent

Barn Swallow

the concrete face of Red Rock Dam. In August I have seen birds stretched over large areas banking insects in the air or resting on utility wires with Tree, Bank, and Rough-winged swallows. Both adults and young are associated with these flocks.

A pretty little bird 5 inches long, the Cliff Swallow may be distinguished by a buffy forehead patch, peach rump, rusty throat, white belly, dark back and wings, a square tail, and short, weak legs and feet. The Barn Swallow has a dark, rusty forehead patch and a forked tail.

Fine architects, the Cliffs combine nest building and courtship. One summer Frances Metcalf and I observed these varied activities at a small concrete bridge near her home north of Pleasantville.

With forty-five mud-jug nests being plastered to the concrete understructure of the bridge, it was a busy, busy scene. A constant pleasant twittering continued throughout the long workday, which was divided between gathering mud pellets, hawking insects in the air, courting, and rest periods on the utility wires with perfect harmony, no fighting.

The gathering of mud pellets required considerable finesse. They fluttered down with feet outstretched, landing gently on the soft black mud. To avoid soiling their feathers, the body and tail were uplifted and to avoid sinking in the mud, the wings, extended straight up over the back, fluttered continually. Quickly gathering a beakful of mud, they rolled it around the mouth to form a pellet, then back to the nest. Since hundreds, perhaps thousands of these pellets go into each nest, it is not surprising that nest construction requires seven to fourteen days. Completed, the nest resembles mud jugs with the jug-neck entrance hole protruding from half an

inch to 5 or 6 inches. The same nest is used year after year with a little repair work done each spring.

After proper drying, the nest is lined with feathers or fine grass. The white eggs sparsely marked with fine brown spots number three to six. Incubation time is thirteen to fourteen days. The young remain in the nest eighteen to twenty-one days and are fed by both parents.

The diet of the swallow species consists entirely of insects. Of all species, certainly the swallows are wholly beneficial with no bad habits, the cheerful, twittering friend of people. One food study reported that 26.8 percent of their food consisted of beetles, 26.2 percent of bugs, 13.9 percent of flies and mosquitoes, with the remainder divided among the other insect orders. McAtee found eighteen southern corn rootworms in one Cliff Swallow stomach, and Beal identified 113 species of beetles in his analysis of Cliff Swallow stomach contents.

Cliff Swallows' colonies are not numerous in Iowa. In Frances Metcalf's study, her questionnaire to Iowa Ornithologists' Union members revealed only eleven localities reporting a total of seventy-four colonies. The greatest number was reported from Greene County by John Faaborg of Jefferson—twenty-nine colonies under bridges and in concrete culverts. Twenty colonies in similar sites were reported from Boone County and seven colonies from Story County. The only colony reported on an old barn was reported by Pete Petersen of Davenport. It was later destroyed by vandals. The other colonies were on rock cliffs.

The rough-lumber barn-eave nesting sites provided by pioneer farmers resulted in a population explosion, with the Cliffs nesting on most of the 1,600 barns then in Marion County—not just one nest to a barn but many, sometimes in the hundreds.

By 1875 or 1880 the population explosion in the House Sparrow was in the making. Subsequently the Cliff Swallow population plummeted as their nests were usurped by the House Sparrows. Only a few octogenarian farmers could recall seeing the swallow nests in barns in their early childhoods.

Swallows nest on rock cliffs throughout North America, from Brooks Range in Alaska south into Mexico. The great concrete dams erected by the Corps of Engineers are fast becoming favorite nesting sites. Seventy-one nests were built on Red Rock Dam in 1969, but many now have been taken over by the messy House Sparrows.

The Cliff Swallows are the famed swallows that supposedly return to San Juan Capistrano the same day in March each year. Not so, say the ornithologists, not always on the same day. In fact, they did not nest on the mission buildings at all in 1964. They arrive in Iowa in May, flying overland from far south in South America through Central America and then fanning out all over North America. They fly only by day, feeding along the way, and they do not fly across the Gulf of Mexico.

Five other swallow species arrive a little earlier. The Tree Swallow, a spring and fall migrant, seldom nests here though the Griffiths found them nesting in bluebird boxes at Brenton's Slough near Des Moines a couple of years ago. Apparently a few nested in the Red Rock Refuge area as there were both adults and young in the late summer feeding flocks of Cliffs this year.

The Rough-winged Swallows nest here in holes in roadcuts, gravel pits, or riverbanks but not in colonies as do the Bank Swallows (known in Europe as Sand Martins), digging their holes with feet and bills. I know one clay bank along the Des Moines River (now within Red Rock Refuge) where Bank Swallows have nested for over fifty years.

The Barn Swallows, a worldwide species, though gregarious are not colony nesters, with seldom as many as five pairs to a barn. Purple Martins are colonial nesters. Both species adapted quickly to humans and now never nest in the wild.

Quoting Roger Tory Peterson (*Birds over America*, 1950), "As a family, the swallows seem to completely be won over to man's way of life. The Hirundinidae are opportunists. They will never allow civilization to displace them." I am not so sure now; although swallows are not yet on the rare and endangered species list, their numbers are reduced since farms and roadsides are saturated with persistent pesticides and herbicides.

A. Cruickshank/Vireo

Cliff Swallow

The return of swallows to Capistrano is a much heralded event in California each spring. These are the famous Cliff Swallows that plaster their mud-jug nests to the walls of the old mission.

For centuries before people reached Iowa these swallows returned each May to nest under the overhanging rocks of Painted Rocks Cliff. Each May I, too, eagerly await the return of these pretty little birds. About sparrow size, they have a buffy forehead patch, peach rump, rusty throat, white belly, dark back and wings, square tail, and short weak legs and feet. The Barn Swallow has a dark rusty forehead patch and a forked tail.

They have adapted easily to humans and now build their nests under concrete bridges, square culverts, and on huge dams. If you want to observe their architectural ability, drive to the tailwater area below Red Rock Dam.

A constant pleasant twittering continues throughout the long workday, which is divided between gathering mud pellets, hawking insects in the air, courting, and resting. All is harmonious, no fighting among members of the large colony.

The adults build their nests with a series of small mud pellets. The gathering of these mud pellets requires considerable finesse. They flutter down with feet outstretched, landing gently on the soft black mud. To avoid soiling their plumage, the body and tail are uptilted. And to avoid sinking in the mud, the wings are extended straight up over the back and are fluttered continuously. Quickly gathering a beakful of mud, they roll it around in the mouth to form a pellet, then fly back to the nest. Since hundreds, perhaps thousands of these pellets go into each nest, it is not surprising that nest construction requires seven to fourteen days. Completed, the nests resemble mud jugs with the jug-neck entrance hole protruding from ½ inch to 3 or 4 inches. The same nest is used year after year with a little repair work. All too often House Sparrows take over the jugs and damage them.

After proper drying, the nest is lined with feathers or fine grass. The four to six white eggs sparsely marked with brown spots are incubated thirteen to fourteen days. The young remain in the nest eighteen to twenty-three days until capable of sustained flight. Both parents feed the nestlings.

Their diet is entirely insects caught in the air. The swallows are a wholly beneficial species with no bad habits, the cheerful twittering friend of people.

Cliff Swallow colonies are not numerous in Iowa but may be found under many bridges, on limestone and sandstone cliffs along rivers and streams, and on outbuildings under eaves.

Carl Kurtz

Gray Jay

Never have I seen a Gray Jay in Iowa and neither has anyone else except a select few in the Decorah area. In November 1976 Darwin Koenig made a positive identification of the bird coming daily to feed at a nearby cabin.

This was the first record of the Gray Jay in Iowa.

A first record for the state requires that two or more experienced birders identify the bird or else a photograph of the bird documents it. A rare bird documentation form must then be completed, signed, and presented to the Iowa Ornithologists' Union compiler of field reports, Dr. Nicholas Halmi of Iowa City. (This is required for most accidentals, too.)

But Nick and Pete Petersen, editor of *Iowa Bird Life*, decided they wanted to observe the bird. They chose a bitter-cold subzero day and nearly froze inside the unheated cabin before the Gray Jay finally flew in while Pete shot a couple of rolls of film. They departed shortly before they reached the frostbite stage.

As a child, I lived two years in northern Wisconsin and later vacationed for many summers in the Itasca State Park, Bemidji, Cass Lake, and Hibbing, Minnesota, areas and in Rocky Mountain National Park in Colorado. Those were my rare sightings of the Whisky Jack, a name given the bird by the early lumberjacks, a corruption of its Indian name, Wiss-ka-chon.

Another common name is Camp Robber, and the name is well deserved. To quote Bent's *Life Histories*, it will eat anything from soap to plug tobacco, at least it steals such items and caches them somewhere.

It will eat any kind of meat, fish, or food left unprotected for a minute and will carry off what it cannot eat. There are many reports of it coming to the camper's table at mealtime and grabbing food from the plate. It is not well liked by trappers, often following them on their rounds stealing bait from traps and damaging trapped animals.

An early ornithologist, William Brewster, told of their pack-rat activities at his camp.

He said they prefer baked beans and oatmeal to meat. He located two of their storehouses: one in a pine stub contained about a pint of biscuits and brown bread and another in a larch stub contained bread tightly packed in three holes, at least a quart.

However, the Gray Jay is largely insectivorous during spring, summer, and fall, feeding on grasshoppers, wasps, bees, and other insects and their larvae. Sometimes it lays up stores of berries in hollow trees for winter, but if food is scarce it has been known to feed on lichens.

Its behavior is characterized by tameness, boldness, thievery, and a big lump of curiosity. It is truly a "Nosy Nell," coming to watch intently any strangers, inspecting them at close range. The sound of an axe attracts them; it means a camp and a camp means food.

Earlier this century the winter logging camps were "winter resort" areas for the Whisky Jacks. My young uncle worked one winter in northern Wisconsin in such a camp where the lumberjacks all slept in a big bunkhouse. Often the Gray Jays were the only birds seen near the bunkhouse and along the logging road for weeks at a time. Our neighbor built the big wooden water wagon used to ice the logging road so that one team of big Belgians could pull the huge sleds full of logs.

Those sleds were much wider than the Iowa bobsleds. A second team was stationed at any grade and hitched on to help pull the sled up the incline and then to hold it back on the decline to prevent it running over the horses and killing them, so great were those loads of logs, the last of the virgin timber.

I have never seen a Gray Jay's nest, as they nest very early, beginning the nest building in mid-February. One observer watched the birds and found that they spent a month at the nest building. It was a very neat, very thick nest 8 to 10 inches in diameter, heavily lined with fur and feathers, so warm that below-zero weather did not freeze the eggs as long as the female brooded them. One observer in Saskatchewan reported temperatures of 32 degrees F below zero during incubation and the eggs did not freeze.

The three to four eggs, grayish or greenish white spotted with different shades of brown, slate gray, and lavender, are incubated sixteen to eighteen days by the female, and the young fed by both parents remain in the nest fifteen days according to Bent.

The Gray Jay is the same size as the Blue Jay but lacks a crest on the head. The forehead and sides of the head and the throat are white, the back of the head is blackish, back, wings, and tail darker gray, and the belly grayish white. Young birds are dull blackish during the summer, but following the post-juvenile molt in July and August old and young are practically alike, although the forehead and back of the young bird may be lightly tinged with brown.

Most campers have seen this bird in the north woods or in the western mountains. If you are a new camper look for the delightful friendly Camp Robber.

Carl Kurtz

Blue Jay

Long ago a distinguished British ornithologist was eager to see a Blue Jay because he considered it the finest bird in the world. Although native to the eastern United States, they have been extending their range westward and we do not consider them an unmixed blessing. Their predation of small birds is too well known.

Of the twenty-seven orders of birds, the passerines or perching birds are by far the largest, containing about 5,100 species. Many consider the finches and sparrows as the most "evolved," but the older ornithologists place the Corvidae—jays and crows—at the top of the family tree. To quote Roger Tory Peterson, "Perhaps they are right; certainly these resourceful birds are plastic, unspecialized, opportunistic, and probably capable of much further evolution and that is what counts."

Big and bold, brassy and sassy, beautiful in its blue, black, and white coat, it is the sentinel of the woods, loudly sounding the alarm, shrieking "jay, jay, jay" if an owl or hawk invades its territory. Pestering owls is great winter sport for the jay, and often in my neighborhood they have led me to a Long-eared or Screech Owl peacefully daydreaming in a cedar tree.

Often I was infuriated by their shrieking at me when I was bird-watching in our Georgia woods. Occasionally they have a shrieking session over the owl and hawk flight-cages at the east side of my house here; I am sure that is done just for fun. With their intelligence, they know full well that the hawks and owls cannot escape. If their shrieking session is prolonged, I get a little annoyed and go outdoors and do a little shrieking of my own. They leave, but I can imagine them saying, "Ha! We got the old lady's attention!"

The jay has many calls and even a rather pretty song. The most amusing is the pump handle or anvil call "tull-ull" in which it raises and lowers its head twice. Much to my surprise, when I investigated a whisper song outside my window in Georgia, I discovered it was coming from a Blue Jay.

They are omnivorous, eating a wide variety of animal and vegetable matter—mice, eggs, nestlings, insects, grain, seeds, acorns, and fruit. All are relished. They are also pack rats, swallowing six to eight sunflower seeds or grains of corn, then carrying them to the storage area.

Many are migratory, moving to southern states to winter, so our feeders are not dominated by them in winter. In Georgia, they were permanent residents, and I banded thirty-one in one month at my feeders before I grew tired and quit.

They are devoted parents, building a big nest of twigs, bark, moss, paper, rags, string, leaves, and dry grasses about 7 to 8 inches in diameter with a 3- to 4-inch cup about 2½ inches deep, usually at a height of 5 to 25 feet, well hidden in tree foliage, often in town.

The four to five eggs may have a pinkish buff or a greenish ground color marked with olive and darker brown spots or fine dots and are incubated seventeen to eighteen days. On hatching they are homely little creatures, entirely naked and blind. The eyes begin to open by the fifth day. On leaving the nest at eighteen to twenty-one days, they wear the handsome Blue Jay plumage but have stubby tails and can shriek as loud as their parents.

Once my mother was wearing a percale sunbonnet while rescuing a baby Blue Jay. Its shrieking brought the parent birds and they attacked. Mother retreated to the house, blood streaming down her face. My advice: wear a hard hat if you must rescue a baby Blue Jay or better still leave the young jays alone!

Carl Kurtz

Clark's Nutcracker

Out of winter's snowstorms come unexpected feathered guests. The December storms of 1972 brought one so far out that it was unbelievable.

As the snowfall diminished during the morning of December 12, my neighbor, Leota Chapman, phoned to report a strange bird at her feeders. This bird, she said, had a gray head and body, heavy white undertail coverts, and black bill, wings, and tail, size of a Blue Jay with jaylike appearance.

I named a few gray birds—Mockingbird, shrike, dove. None of those. She further stated that it was not in Peterson's *Guide to the Birds East of the Rockies*. Facetiously I said, "Clark's Nutcracker"—pure wild guesswork, since the last authentic Iowa record was in 1894. These birds live in the high Rockies just below the timberline.

Bill and Gladys Cummings, second house south, had also seen a strange gray-and-black bird at their suet feeder a couple of days earlier.

For the next three bitter-cold subzero days I took a daily walk around the block but found no strange bird. Finally my luck changed. At 2:25 P.M. on December 15, I looked up when a bird landed with a heavy thud on the east window shelf. I couldn't believe my eyes—it was a Clark's Nutcracker! It hammered with powerful blows at the frozen suet then flew away with an acorn-sized chunk. Thereafter it was a daily visitor at my feeders.

Later the librarian, Marion Batten, informed me the bird had been eating at her feeders since December 1. She had dubbed it the "mystery" bird because it was not pictured in the *Eastern Bird Guide*. I was indignant—a bird so rare that it hadn't been seen in Iowa in seventy-eight years and Marion hadn't called me.

Of all the rare birds, the accidentals are the rarest of the rarities—birds that should not be in a region at all. Most ornithologists discover few in a lifetime. I have been fortunate in discovering two, the Western Tanager in Georgia in 1956, the second recorded for the state and authenticated by Dr. David Johnston, and this nutcracker sight record which had to be authenticated either by another ornithologist or by photographs.

I called a fellow member of the Iowa Ornithologists' Union, Herb Dorow of Newton, a wildlife photographer, but no bird showed on the mornings that Herb came. Refusing to accept defeat, Herb set up one of his cameras on my tripod in front of the east window, loaded it, focused it, and instructed me to shoot a full roll—twenty exposures. This I did the next day, and much to my surprise, several of the photographs were good although most were not.

When Lewis and Clark in 1805 made their famous trip over the Rockies and down the Columbia, they discovered a woodpecker that looked and flew like a corvid (crow family) and a corvid that looked and flew like a woodpecker, with undulating flight.

The first was named after Lewis—"Lewis" Woodpecker—and the second after Clark—"Clark" Nutcracker or woodpecker. It was not a woodpecker; it is a nutcracker, although it does indeed fly like a woodpecker on short flights between trees (but not on long flights, I noticed).

According to Bent's *Life Histories*, the nest and eggs were not described until March 1876, when Major C. E. Bendire, wading through 2-to-4-foot-deep snow in the high Rockies, found a round ball of a nest perched on a horizontal pine branch. It was constructed of sticks, bark strips, grass, and pine straw, necessarily a snug, warm nest.

The two to three eggs, rarely four to six, are greenish gray, finely sprinkled with dark brown and lavender, and require seventeen days incubation. The nests are found only among pines bearing ripe cones because the nestlings are fed exclusively on hulled pine seeds. Fledged at eighteen days, they fly after the parent birds, begging for food and squawking at the top of their lungs.

Omnivorous, the adult's diet includes anything edible—piñon pine seeds, cedar and juniper berries, seeds, acorns, nuts, fruits, insects, small mice, and even carrion. They accept handouts at camps, becoming quite tame. Like others of their family, they are nest robbers, eating eggs and small nestlings.

This is a delightful bird to watch—such a busy, bouncy individual. One afternoon a Clark's Nutcracker extracted a piece of suet the size of an apple from the feeder and glided with it to the snow about 20 feet from my north window. The bird hammered off several cherry-sized pieces and swallowed them, then hopped a few feet into neighbor George's raspberry patch.

It wriggled its neck and regurgitated one suet chunk, hammered it into the snow, then repeated this two more times, placing all three chunks in the same hole. It picked up chunks of snow in its beak, filled up the hole, and smoothed it over with a few sweeping motions of its beak.

Thereafter it carried a single suet chunk in its beak, burying each in a separate hole. When it had reduced the big chunk to half size, it buried the entire piece in the snow, then flew northeast to the neighbor's pecan tree, pausing briefly before flying into a big cedar tree.

In less than an hour it returned, switching its "Operation Deep Freeze" activities to the east window shelf, carrying off several chunks of suet and burying them in the snow under the apple tree.

After it had driven one piece of suet into the snow, I saw it pull a devious trick. It picked up an empty walnut shell and tamped it into the hole on top of the suet, then covered all with snow. Was it to thwart future robbers or was it just busybody activity? In the cold, cruel, harsh world of the Hudsonian life-zone, these birds have learned all the tricks of survival.

Jays, crows, magpies, and nutcrackers (113 species worldwide) belong to the corvid family of the order of passerine (perching) birds. Roger Tory Peterson states that "European ornithologists usually place the intelligent crows and jays at the top of the family tree whereas in America we rate finches highest. These resourceful birds are relatively unspecialized, opportunistic, and probably capable of much further evolution—and that is what counts." I agree fully with the European ornithologists.

The appearance of the nutcracker in Pleasantville during December and a report of one near Ottumwa, also three in Missouri, may be indications of a widespread irruption similar to the one in 1894.

Irruption migrations are extraordinary north-south migrations triggered by hormonal changes. They occur when permanent residents are forced to migrate because of food shortage, in this case, failure of the cone crop in the mountain regions.

Watch for this big, boldly marked bird in Iowa. The black tail has white outer feathers, and the black wings have a large white spot in the secondaries. The head and body are gray, and the undertail coverts are white.

Carl Kurtz

Black-capped Chickadee

The chickadee has been called "the bird of the merry heart" and that is no exaggeration. Day after dreary winter day this tiny sprite brightens my east window shelf, not once but many times. Coming in rather slowly, it lands with a tiny thud. Sometimes it sorts through the sunflower seeds, but more often it grabs a seed and is gone in a flash. With very rapid wingbeats, about thirty-five per second, it can turn course in three-hundredths of a second, literally a flash!

Never are two on the shelf at the same time; one will wait its turn on a swaying apple twig or dart around the house to either the north or south feeder. They are extremely gregarious, never appearing alone except during the nesting season.

There are no longer numerous big flocks. The loss of elms, the clearing of the trees along farm ditches, the change to iron fence-posts, and pesticides have reduced the population drastically.

When I come out to replenish the feeders or remove a bird from a trap, I am soundly scolded until I return indoors, then, triumphant at having vanquished the big enemy, they resume feeding.

When caught in a mist net, they never cease wiggling, becoming so entangled in a few minutes that I often cut the net rather than risk hurting them. I much prefer capturing them in traps during the winter. When Mary Felsing and I are using the nets at her home—where there is a huge chickadee population—we watch constantly to free an entangled chickadee. No other species is such a hyperactive ball of fury in the net.

The "chick-a-dee dee" is the call best known, but they actually have many other calls. The next best known is the "fee-bee" call. Georgia's noted ornithologist-ecologist Eugene Odum, in his 1940 studies, described sixteen different calls given by the Black-capped. The "si-si-si-si" warning notes, usually reserved for hawks, were sounded the winter there was a Long-eared Owl in the neighbor's cedar.

The only pattern I ever have detected in the courtship ritual is a frenzy of hyperactivity. One venerable ornithologist described it nicely: "The birds grow agitated late in March and increase their vivacity during April." My observation has been that the courtship is rather short, with the attention soon focused on a nest site, any small cavity.

They accept bluebird nest boxes readily, especially if placed near brush or timber, nesting quite early before the arrival of the predatory little House Wren. Heaven help a late-nesting chickadee. The sharp little bill of the wren punctures each egg as neatly as if done with a hypodermic needle.

I recall very well the first time I looked into a bluebird box containing a mass of fur. I reached in to clean it out, sure that it was a white-footed mouse nest. Luckily, I felt a tiny egg, and then I learned that the chickadee covers her eggs until the full clutch is laid.

The base of the nest was of moss and contained rabbit, raccoon, and fox fur.

The six to eight tiny white eggs dotted with reddish brown are incubated about twelve days, with the male feeding the female. Both parents feed the young and often become very thin from overwork.

Once in 1963 Ruth Summy Collins and I were doing nesting studies and located a chickadee nest in a cavity of a big red haw tree. Ruth's hand was small enough to reach the nestlings. One by one she brought them out, and I banded them, a total of eight plump little chickadees. That was during the last big irruption of elm inchworms. Soon the elms died, and I never again found a nest of eight chickadee nestlings.

The same year Ruth and I located another nest in an old fencepost. It was a cool day and the little mother refused to leave her babies, so Ruth reached in, picked her up, and I banded her and seven plump nestlings. We were scolded soundly all the while.

The probable length of life is two to three years, but there are records of banded chickadees living seven and one-half and nine years. The oldest ones I have ever retrapped were three years old. Chickadees have a maddening habit of repeating, getting in the trap a number of times. Little Band No. 10 was a character, in the trap at least once a week all winter long, and each time just as mad as it could be, biting furiously at my fingers.

They are highly acrobatic, and I many times have seen them upside down under a limb gleaning insects, which comprise 70 percent of their diet.

Of the seven chickadee species in North America, the Black-capped, ranging across all the northern states into southern Canada and coastal Alaska, is the tamest, coming daily to feeders, whereas the Carolina Chickadee of the southeastern states is much more timid and seldom came to our feeders in Georgia.

The Crow, the Northern Cheyenne, and the Blackfoot Indians believed the chickadee was a wise bird, "least in strength, but strongest in mind among its kind," able to withstand bitter subzero cold and the most violent winds in the mountains, sleeping at night in snug tree cavities.

Larry Stone

Tufted Titmouse

One winter my days were brightened by three bouncy little Tufted Titmice and their pack rat activities, coming to the feeders off and on all day long. How did I know I had only three? Because each wore a shiny new aluminum leg band I applied early in the winter when their nosy traits led them to investigate the traps. In the hand they appeared to be more angry than frightened, biting furiously at my fingers.

As a child I never saw a titmouse, and through the years I have never had more than three at my Iowa feeders. A Carolinian species of the South, they were far more numerous at our Georgia feeders. Like the Cardinal and the Red-bellied Woodpecker, they gradually extended their range northward following the pioneers.

In 1907 R. M. Anderson ("The Birds of Iowa") rated it to be a rare resident in southern Iowa, seldom reaching the northern part of the state. By 1933 Phil DuMont (*A Revised List of the Birds of Iowa*) considered it a common permanent resident in the south but fairly rare in the north. Woodward H. Brown in 1971 ("An Annotated List of the Birds of Iowa") rated it a common permanent resident in Iowa.

The largest of the four North American titmice and the only one reaching Iowa, it is a trim, slender, sparrow-sized bird of soft gray with a crested head, white cheek, breast, and belly, and peachy sides. The black beady eye is conspicuous in the white cheek patch. Just above the black beak there is a small black square.

The cheeriest of all sounds on a gray winter day is the "peto, peto" whistle of this bird. While most species are silent through the winter, these delightful little sprites are more often heard than seen in the winter woods of Red Rock Refuge, around Red Rock Lake, and in town.

They mate for life, remaining together throughout the year, with the male definitely dominant. Even though my east window feeder is 3 feet wide and 1 foot deep, the male would not allow his mate to eat with him. She had to wait in line, a situation I found intolerable.

I fixed that little male chauvinist pig. I added three more sunflower-seed feeders, one at the north window, one at the south, and the old weathered board shelf with the rusty tuna can was brought into use again at the other east window. The waiting in line on bitter cold mornings ceased.

But when it comes to collecting materials for the nest, no bird is more aggressive than the female titmouse. Fur is highly prized, not just any old fur but nice fresh fur from living mammals, as well as human hair.

Numerous records are scattered through the literature of their plucking hairs from the tail of squirrels, from the back of a woodchuck, or from other animals. In Des Moines, Woodward H. Brown once watched one vigorously yanking bristles from the back of a hog swimming for dear life in the floodwaters of the Raccoon River.

A South Carolina lady watched a cat clinging precariously to the top branch of a tall tree while a titmouse plucked a beakful of fur from its tail. At least two human birdwatchers sitting quietly have had hair snipped from their heads by a titmouse. Cavity nesters, they often use an abandoned woodpecker nest hole, rarely a bird box. The four to seven (seldom eight) white eggs are dotted or spotted with brown and incubated thirteen to fourteen days by the female.

The nestlings are fed by both parents for seventeen to eighteen days on insects. The young birds stay with the parents in a small family flock through the first winter. Occasionally, a yearling without a mate becomes a "helper," assisting the two parent birds in feeding the nestlings.

They are utterly fearless of humans. I once stood 20 feet away from the nest tree in a busy picnic area of Springbrook Park while we humans were completely ignored!

Their diet is about 66 percent animal

matter, insects and larvae, and 33 percent vegetable—mulberries, Juneberries, Virginia creeper, and so on. Acorns are a staple of the winter diet.

They were absent from my feeders for a couple of winters until I recalled the words of the late Murl Jones (Iowa's veteran bird bander of forty years) that the aroma of black walnut meats was irresistible to many birds. Sure enough, the titmice came for the walnuts and remained to eat sunflower seeds all winter.

Bird behavior is fascinating to study. This winter for the first time I observed one of my titmice "shadow boxing" its reflection in the window pane. Unlike the Cardinal it did not peck the glass but it did glare angrily at its own image, opening its beak wide and hissing at this invader of its winter territory.

Larry Stone

Red-breasted Nuthatch

Red-breasted Nuthatches are great pack rats. One November morning I watched a little Reddi carry sunflower seeds or bits of suet, one at a time, and stash them in the bark of nearby trees or in holes of old posts. A bundle of energy, she also planted five seeds in bare ground, unusual behavior that I had never before observed. For a couple of hours at noon she disappeared, probably for a siesta. Goodness knows she had earned a rest! Before the day ended she had discovered all five sunflower-seed feeders and had "pack-ratted" seeds from each one.

Our permanent resident White-breasted Nuthatch is larger, about 5 inches long, and much slower moving than the 4-inch-long Redbreast which has a black cap, blue-gray back, wings, and tail markings, similar to the Whitebreast, but it has a white stripe above a black stripe through the eye. The male has a rusty red breast but the female's is more of a rusty yellow.

Aside from North America, this species is found in only two other places in the world—in far-off Corsica and in North China. Their natural habitat is the coniferous forests in the mountains of the far north, where they feed on the seeds they extract from cones with their long, slender bills.

Failure of the cone crop produces those southward movements known as massive invasions. Sometimes population pressure also causes the lemminglike wanderings of this little bird; flocks have been seen flying out over the Atlantic Ocean to certain death from exhaustion and drowning. The winters of 1968–69 and 1969–70 were massive invasion years which was surprising, one invasion right after another.

Over the past twenty-one years I have had a Redbreast spend the winter with me only twice. Like many birds of the far north, they are very tame, and one would take walnut meats from my hand. Occasionally one appears, "pack rats" madly for a few days or a couple of weeks, then moves on. I'm hoping this little Reddi stays, so delightful to watch.

There is only one nest record of a Red-breast in Iowa, reported by Mary Ellen Warters, May 1958, in a cavity at the top of a 7-foot post. The full clutch is four to seven white eggs finely dotted with reddish brown.

Larry Stone

Brown Creeper

The little Brown Creeper is a solitary bird of the woodland never seen in flocks and an uncommon winter resident in Iowa. On Christmas Bird Counts throughout the state the number varies from two to eleven per count area, a 15-mile-diameter circle.

About the size of a House Wren, it resembles a dead leaf, with brown-streaked plumage on the head and back, a white eyeline, and white breast and belly. Its bill, nearly a half-inch long, is slender, pointed, and down-curved, ideal for probing insects from tiny crevices in tree bark. The tail is brown, rather long, pointed, and stiff like a woodpecker's tail, an aid in its climbing search for food. The toenails are very long, another aid in climbing.

It is always a joy to watch one feeding in the woods or in my yard. Invariably, I follow it from tree to tree in the Red Rock Refuge woods until it flies too far away to follow. Fascinated, I watch as it flies to the base of a tree, then spirals around and around with hitching movements, climbing higher and higher, constantly probing and picking out insects and eggs. But it does not hammer like a woodpecker.

Occasionally it takes a backward hop or one sideways to snatch an insect too tiny for me to see. When the tree has been thoroughly searched to the top, it flies diagonally downward to the base of the next tree and begins the spiral staircase climb all over again.

In the middle of winter, a kind lady, Mrs. Albert Dunkin, brought me a Brown Creeper picked up on her porch floor that bitter sub-zero morning. Since the bird had no apparent injury, it probably was a victim of exposure, forced out of its winter cavity roost for some reason. Most of our small winter birds such as the chickadee and nuthatch survive by sleeping in tree cavities well protected from the severe cold.

Such a small bird easily could escape through the bars of a birdcage, so I substituted a long glass terrarium with a screen lid. I placed an old bark-covered slab birdhouse at one end and a bark-covered slab at the other end after rubbing suet and peanut butter into the rough bark. At first, the little creeper searched for a way of escape, made a few jumps up the glass sides, then began a spiral climb around the old birdhouse, probing and eating bits of suet. Finally, it entered the hole of the box and stayed inside for a minute or two.

Over the years I have been interested in the time of rising and of going to roost and have found that it varies with different species. The Cardinal is almost always the first bird at the feeder each morning and the last to retire in the evening. The chickadee usually arrives about five minutes later each morning. On New Year's morning the first Cardinal arrived at the feeder at 7:10 A.M. and the first chickadee at 7:15 A.M.

So I decided to watch closely to see when the Brown Creeper would go to roost and where. Sunset that day was at 4:54 P.M. central standard time. At 4:28 P.M. the creeper entered the bluebird box through the hole and made the characteristic "zit-zit" calls off and on for ten minutes. Thereafter no sound at all. When I peeked in the little bird was on the floor of the box with its body pressed against one corner and its beak tucked under its right wing.

I celebrated via television until 1:00 A.M. I did not place a dark cover over the terrarium as I usually do with a cage, but, undisturbed by lights and noise, the bird slept on. Each time I peeked it was sleeping peacefully.

Next morning I planned to band and release it after a period of feeding, but first I was curious as to when it would rise. Sunrise was 7:41 A.M. The Brown Creeper emerged at 7:37 A.M., a bundle of nervous energy, "zit-zitting" now and then, spiraling the old box, and picking bits of suet from the bark.

Occasionally alighting on the screen cover, it walked upside down with the greatest of ease, sometimes in a straight line and sometimes in a circle. This had been my only chance to observe a Brown Creeper in captivity, as the few I have banded were released immediately as required by the Banding Office, U.S. Fish and Wildlife Service. (Banders are permitted to hold birds briefly for educational purposes only.)

The Brown is the only member of the creeper family found in the United States. (It also is widely distributed in Europe.) It breeds in the northern states to Alaska, in Canada, and in the mountains of our southern states, wintering all the way to the Gulf and as far south as Nicaragua.

The creeper's attachment to tree bark carries over to its nesting, for the hammock-shaped nest is placed behind a piece of loose bark from 5 to 30 feet up. During courtship, the male forgets he's a creeper and becomes an aerial sprite, madly flying spirals around tree trunks.

The five to eight white eggs are finely dotted with reddish to dark brown spots. The nestlings leave the nest at thirteen to fourteen days, simply moving out on the tree bark to begin their creeper lessons. It rarely ever nests in Iowa, but on June 12, 1968, John Faaborg observed an adult feeding young in Boone County.

A. Cruickshank/Vireo

Carolina Wren

Every bird-watcher is eternally in search of a rare bird so it was a real thrill to pick up the phone on June 8, 1973, and hear Mary Felsing say quite calmly, "We have a pair of Carolina Wrens nesting in our garage." My reply was slightly hysterical. To have a rare bird nesting in a residential garage is an exciting happening, and I asked many questions.

The Felsings were only mildly excited—they have more bird species at the feeders of their beautiful home in the woods near the south end of Red Rock Dam than I have ever seen at any other feeder in Iowa. I always overstay my welcome there.

Mary reported the male to be a beautiful singer, loud and persistent, outside her kitchen window. That I know to be very true, as we had them every day at our suet feeders in Georgia and they sang most of the year.

This bird has been called the Mocking Wren because of the variation and versatility of its songs. The most common songs are "tea-kettle, tea-kettle, tea-kettle" and "jo-reaper, jo-reaper, jo-ree." It does an excellent Cardinal whistle and another that sounds like the Tufted Titmouse's "peto, peto, peto."

Even after hearing these birds for years, I was still occasionally fooled by their variations.

A southern species, rated a rare permanent resident in Iowa, they have been extending their range northward throughout this century. The population increases until a severe winter decimates the majority. After the heavy snows of the winter of 1961–62 I did not see a Carolina Wren again until I banded one on November 10, 1965, trapped just outside my door.

Of the six wrens found in Iowa—House, Winter, Bewick's, Long-billed Marsh, and Short-billed Marsh, the Carolina is the largest, 5½ to 6 inches in length, and the brightest, with a reddish brown head, back, and tail, a white throat, rusty sides and belly, and a conspicuous white eye stripe. The Carolina is a staunch defender of its territory. Mary said this one delighted in harassing the Downy Woodpeckers at the suet and was not too kind to the House Wrens feeding babies in their nest box just outside the garage.

The nest was typical, big and sloppy with leaves, sticks, and roots in an open box atop a rafter. Nests in the open are domed over with a small entry hole in one side, and this was true of the nest at the Felsings. The Carolina almost always builds in woodlands or thickets under brush piles, uprooted trees, or in tree or stump cavities, sometimes in crotches of trees.

Building "dummy" nests is a wren trait shared by the Carolina. My husband and I once had the pleasure of watching a male spend most of a morning picking up big oak and gum leaves and filling the space around the feet of Saint Francis in our garden shrine box. The snooty little female came, inspected the nest, took a good look at the many birds at our feeding stations, and departed as if to say, "You foolish male, this is no fit place to raise a family." We sympathized with the hardworking little male, but he was undaunted, and singing merrily, he built the next nest on a sill under a neighbor's house, and that one was accepted.

Four to six, sometimes eight pinkish white eggs with brown spots is the usual clutch. Mary reported the parents carrying food on July 1 and July 11—all had made successful flights. Often the father takes charge of the first brood of fledglings, teaching them to hunt food while the mother begins incubating the second clutch of eggs, but Mary reported there was no second brood in the old nest.

Carl Kurtz

House Wren

In 1925 an able Iowa ornithologist, Althea Sherman, of northeast Iowa took a venomous stand against the House Wren in a spirited article, "Down with the House Wren Boxes." It is true that the House Wren is a great destroyer of nests and eggs, neatly puncturing the eggs with hypodermic holes made by the sharp bill. The wrens have never heard of birth control, so they persistently lay six to eight pretty pinkish eggs and have remarkable success in rearing their babies too. I once found a nest of ten eggs!

The puncturing of other birds' eggs is done to protect the food supply within their nesting territory. This behavior should not be condemned by our human standards. We have upset the balance of nature with our interference by erecting too many nest boxes. One box per large lot should be the limit. The aggressive behavior is clearly evidence of its superior intelligence in the battle of the survival of the fittest.

Many a time I have gnashed my teeth in anger when I opened a bluebird box to find the wren eggs or bluebird eggs all neatly punctured. Other nearby open nesters such as the Yellow Warbler, Yellowthroat, Indigo Bunting, Chipping Sparrow, and Bell's Vireo suffer the same fate, and sometimes a Downy Woodpecker's nest hole is invaded and depredated.

Each May I nail an old wren box to the frame of the north window of my favorite bird-watching room, which has windows on three sides. Then I watch the male, the first to arrive, fill up the box with sticks. It is a joyful occasion when he brings the female on a tour of his nest box. When she accepts the box, he is ecstatic, quivering his wings with joy. Now it is her turn, and she builds the small round nest for the eggs, using soft plant materials.

After an incubation period of thirteen days the diligent, devoted parents feed the young frequently. On July 1, 1972, I timed the feeding trips with a stopwatch. Beginning at 6:45 A.M. and continuing through 8:00 P.M., insects were brought at intervals of 46, 27, 37, 33, 50, 46, 29, 58, and 40 seconds, with the longest interval being 5½ minutes. The time each parent spent in the nest stuffing insects down each throat varied from 7½, 6½, 7, 9, 5, 4½ to 3 seconds.

The wrens may be ornery little cusses, but they are worth their weight in gold as destroyers of harmful insects. At least 98 percent of their food is insects.

I know of no more welcome sound than the cheerful song of the House Wren. The Indian name for the wren was O-du-na-mis-sug-ud-da-we-shi, meaning a big noise for its size. Furthermore, the male sings often not only during courtship but also while nestlings are being fed.

A. Cruickshank/Vireo

Winter Wren

If you see a wren in winter, do not fear for your sanity. It is the Winter Wren, a rare winter visitor in Iowa. Of the ten North American wrens, only six reach Iowa.

Tiny, hyperactive birds with an extremely high metabolic rate, this one is the smallest, only 3¼ inches long. The House Wren is a full inch longer. All others are summer residents except the permanent resident Carolina Wren, a southern (Carolinian) species which has extended its range northward only to be decimated by a severe winter of deep snows.

I am sure to be criticized as unscientific

when I say the Winter Wren is the cutest of all the wrens, very dark brown with dark brown barring on the belly, a light tan eye stripe, a very short, upright tail, and almost continuous bobbing action.

To come upon one of these shy, secretive little birds, usually along ditches strewn with debris driftwood and brush piles, is pure pleasure. By standing perfectly still, I have found them to be very inquisitive people-watchers, darting in and out of brush piles looking me over.

During part of one winter I saw one often as I walked my dog here in town. We were intruders as we walked along the ditch in the acreage, and the wren seemed to be escorting us, keeping just ahead, maneuvering through the brush and weeds, through the swordlike leaves of the bed of sweet flag (calamus), and past the dump of broken cement blocks.

There I held my breath because a mink had its den under the blocks. Once I came upon the mink trying to reach a junco entangled in my mist net loaded with small birds. (Fortunately, the net stretched across the ditch was high and tight so the netted birds were safe.)

I seldom saw the bird in open flight, but occasionally it would dart into the old barn, through the door, and out an open window at the far end. One day I saw it fly straight at an old slatted corncrib of native lumber, through a crack, and out through a crack on the opposite side with no slowing, no hesitation. As navigators through all sorts of cracks, crevices, and tiny openings in seemingly inaccessible places, they are indeed highly skilled.

I watched this delightful sprite nearly every day for a month, then one bitter cold day it was missing. I hoped it had moved south and had not fallen victim to the mink or the bitter cold.

Most do migrate farther south except during mild winters, but many have been found frozen or starved to death during severe weather. The greatest numbers on the Christmas Bird Counts reported in the National Audubon Society's *American Birds* are found in the southern states—17 at Lake Georgia-Pacific, Arkansas, 21 at Shreveport, Louisiana, 21 at Charleston, South Carolina, and 15 at Athens, Georgia. But it is in the cool moist forests of the northwest that the population is highest—96 at Bellingham, Washington, 80 at Gray's Harbor, Washington, and 116 at Cottage Grove, Oregon. Our Iowa count was only 3 in 1968, 7 in 1971, and 21 in 1972, with twenty-seven stations throughout the state reporting.

One birder in western Washington watched 31 Winter Wrens enter a 6-inch square bird box to spend the night during a severe cold spell. The 31st bird gained entry on its third attempt, after many others had failed to get in.

The breeding range extends from southern Alaska east to Newfoundland and across the upper third of our northern states, with some nesting in the mountains of the southeastern states as far south as Brasstown Bald in northern Georgia.

They prefer dense forests and are one of the few birds to be found in the deep dark forests of western Washington, perfectly at home in the dense ground cover of ferns, vines, briers, and mosses ranging from sea level to 5,000 feet in the Olympic mountains.

The courtship consists of much wing quivering and wing "fanning" so that the few white markings on the rump are conspicuously displayed. I have never heard one sing, but others have reported that "the song consists of warbles, rapid notes, and trills interspersed in a great variety of ways, the most beautiful of all wren songs." The alarm note "cheet" is somewhat similar to that of the Song Sparrow.

The nest, built near ground level of twigs and mosses lined with fur or feathers, is most often placed in upturned roots, cracks, or behind the loosened bark of fallen trees. The four to seven, sometimes ten eggs are white, sprinkled with fine reddish brown dots. Incubation of about fourteen days is the duty of the female, while the nearby male serenades her with glorious song.

These entirely insectivorous birds feed their young every one, two, three, or five minutes from early morning til dark of evening. Seldom are feedings as far apart as ten minutes. They are fledged at about nineteen days then travel as a family, the fledglings begging constantly with a soft "zee" note.

Take a rugged hike along a messy ditch in December or January of a mild winter or in late March and you may be richly rewarded by the antics of this tiny wren.

Carl Kurtz

Eastern Bluebird

Our beloved Eastern Bluebird has been described as carrying on his back the blue of heaven and on his breast the rich red of Georgia earth. It was my pleasure to observe the bluebirds on their winter grounds in Georgia over many years. And there I had my "bluebird days" when I would drive to the country looking for a flock in grassy fields with scattered trees and shrubs. Locating a flock, I would park and spend hours observing their feeding habits.

Sometimes they hawked flying insects flycatcher style or they darted among the leaves of green bay or other broadleaf evergreen trees and shrubs. Most of the insects, however, were taken on the ground, the bluebird fluttering down from a low perch, sometimes hovering a few seconds before taking the insect and then returning to the perch to eat it. Although no true songs were heard in winter, the plaintive call notes "oola" or "aloala" added music to those scenes that are etched forever in my memory.

I knew the bluebird from my earliest childhood on the Iowa prairie when they were plentiful, nesting in old fenceposts or in cavities of the old trees or in orchards. Today the bluebird is in serious trouble. Iron posts have replaced the wooden posts, timberland is fast disappearing, bulldozed to make way for row crops, ecological imbalance continues from overuse of herbicides and pesticides, and many old pastures have been plowed up.

This species has plenty of natural predators too. Cats, raccoons, squirrels, snakes, and Blue Jays raid the nests, killing the female and taking the eggs or young. Tiny House Wrens may puncture the eggs, while House Sparrows and Starlings often usurp nest cavities.

In my childhood, the bluebird had no human enemies. Today it is a sad story when vandals destroy nest boxes and hunters use nest boxes for target practice, reducing them to splinters.

Extremely cold weather on their wintering grounds or on their nesting grounds following spring migration kills thousands. Our April 9, 1973, blizzard very nearly wiped out the Iowa bluebird. Few were seen that summer on bluebird trails, and the population is still very low. The severe winter of 1894–95 in the southern states where everything was icebound or snowbound to the Gulf of Mexico killed hundreds of thousands of bluebirds and Robins.

Some thirteen years ago Scoutmaster Lester Hancock and the Boy Scouts of Troop 363 built over one hundred bluebird boxes using bark-covered slabs from a sawmill. This proved to be one of the most delightful and rewarding experiences of my life. The boys gained much, too, and one is now a wildlife biologist.

It was also work, as a bluebird trail requires considerable care. I covered one-fourth of the township each week, timing my visits between 5:30 and 9:00 A.M. during hot weather. Nest records were kept on every brood from nest construction, laying of the four to six pale blue eggs, through incubation, to the banding of the nestlings in the pin-feather stage, to fledging. Each pair nested twice beginning in late March or early April, and if a nest was depredated, it meant a third nesting attempt, with the last nestlings banded in August.

There was a constant battle with the House Sparrows, each time we tore out their nests and destroyed them. The white-footed mice also usurped boxes, and I removed their nests also until the mother mouse had her babies, then I simply didn't have the heart to evict them. After all, they live mainly on weed seeds and waste grain. House Wrens return too late to destroy first clutches of eggs, but they took their toll of later clutches.

One summer, I banded 142 bluebird nestlings, 85 wren nestlings, and 25 chickadee babies.

Yes, a bluebird trail is a worthwhile project, and if we expect to save the bluebirds, we must establish bluebird trails all over Iowa. This is a fine project for Boy Scout and other youth groups. Put up the boxes by mid-March and take them down before September 1, before squirrel hunting begins. Clean, repair, and store them during the winter.

Following are directions prepared by Troop 363 of Pleasantville for constructing the boxes and establishing the trail.

Boxes: Floor—4 × 4 inches; depth—10

inches; 1½-inch hole, 6 to 8 inches above floor, with lid pivoted on a nail so it can be swung sideways to allow observation of the nest. Another nail is inserted in a hole drilled through the top and into a side panel to lock the lid. The back is extended below the box to provide a place to fasten it to the post. The front and back panels are cut a quarter-inch short to allow a little ventilation space under the roof, and the corners of the floor are rounded off just enough to allow drainage at the bottom.

Trails: A bluebird trail consists of ten or more nest boxes erected on fenceposts or other suitable support, 3 to 6 feet above the ground, spaced not less than 500 feet apart along a route that can be conveniently visited along a country road. Be sure to obtain permission from the landowner. The best location is near a bluegrass pasture, away from thickly wooded fencerows or brushy areas. Avoid intensely cultivated fields with no grasslands.

Carl Kurtz

American Robin

The Robin, our largest thrush, is the favorite yard bird in Iowa and other northern states, just as the Mockingbird is the favorite in the South. My mother always watched so eagerly each spring for the first Robin, and I find myself doing the same. In the midst of a February snowstorm, I saw my first Robin chirping in the top of the old Anoka apple tree. It is a true harbinger of the coming spring, so welcome after our long winters.

In Georgia, I enjoyed them all winter long, often watching the flocks flying to their roosts in inaccessible areas of the Ocmulgee Swamp. However, they did not nest in Georgia; they returned north in spring.

How we welcome the Robin's "cheerily, cheer" song. Although they sing off and on throughout the day, it is in the long morning and evening choruses that they excel. A few males begin to sing in darkness and, as the dawn approaches, more and more birds join in until the calls swell to a glorious Te Deum Laudamus. Their evening chorus always reminds me of the beautiful canticle in the Apocrypha, Benedicite Omnia Opera Domini, a joyous song of praise.

Robin courtship is a conspicuous affair with considerable chasing. Once the pair bond is established, the male struts around the female with tail spread and wings shaking.

Once the nest site is selected, they become quite territorial. Both work at building the substantial nest with coarse stalks, twigs, bits of paper, and string topped with a deep mud cup. The female does most of the work forming the cup and lining it with fine dry grasses.

Most nests are located in trees, but I have seen them in some odd sites—window ledges, gate posts, drain pipes, beams inside old sheds, porches, trellises, and meter boxes. One containing a very large clutch of seven eggs was placed in a revolving bird feeder at our Methodist parsonage. The spring winds blew and blew and the feeder whirled round and round like a carousel, but the Robin faithfully incubated day after dizzy day.

A nest that I studied intensively was located in my apple tree, and I climbed the stepladder to check every other day. The Robins ignored everyone else, but the minute I stepped out of my door I was greeted by hysterical scolding which continued until I retreated.

A severe windstorm blew this nest to the ground, breaking one egg. Quickly I climbed the ladder and secured the nest to the same site, using fine copper wire. The female immediately resumed incubating and the three eggs hatched. Did I earn a little gratitude? No, indeed! My every appearance brought on a fit of hysterics.

Four eggs of "robin's-egg blue" is the usual clutch, and the incubation period is twelve to fourteen days. On hatching, the babies are mostly naked, with patches of gray down on the head, and the parent birds stuff them all day long for fourteen to sixteen days.

On leaving the nest, the young birds' breasts are speckled, a characteristic of all thrush babies, and their tails are stubby. This is a time of high mortality as they get down on the ground and often will not stay in a shrub.

Once I was trying to save a very stupid young Robin. I placed it in a chicken-wire trap with a can of angleworms alongside. As I watched, the parent bird came, picked up a worm, and fed little dummy. Then she picked up another worm and flew off with it. I fumed at this act of selfishness, but as I passed my north window, I saw her feed another fledgling perched on top of a raspberry stake. I had misjudged this devoted parent.

The Robin's diet is 60 percent vegetable and 40 percent animal matter. Around our home, they prefer strawberries and cherries. They eat almost all kinds of wild fruits—mulberry, wild cherry, berries, sumac, and Virginia creeper. In winter they relish the blue berries on red cedar trees. On our 1977 Christmas Bird Count we found a flock of nineteen Robins feeding on cedar berries at a deserted farmstead. In fall they feed heavily on the berries of my neighbor's old hackberry tree, along with the Cedar Waxwings. Everyone knows Robins love angleworms, but they also take a wide variety of insects, millipedes, sow bugs, and snails.

Linda Kurtz

Catbird

This bird is almost universally known as the Catbird, but in the south it is sometimes called the Black Mockingbird. In Bermuda, where there are no true blackbirds (Icterids), it has been named the Blackbird.

It is the most abundant of the three mimic thrushes reaching Iowa each summer. A plain-colored, very neat bird, it is charcoal gray all over except for the black cap on the head and the rusty undertail coverts and is about Cardinal-sized.

A fine singer, perched in the tallest shrub in its tangled thicket of bushes and vines, it sings snatches of all the bird songs in the neighborhood, repeating each phrase only once. (The Brown Thrasher repeats each phrase twice and the Mockingbird three to four times.) Sometimes it has the quality of a ventriloquist, difficult to trace to its source; again it is loud and easily traced. It not only mimics bird songs, it may mimic a cackling hen, a creaking wagon wheel, or the croaking of frogs. At least two observers taught wild Catbirds to imitate whistled calls and "whip-poor-will." One was observed doing a typical courtship flight song of the Bobolink, flying upward 30 feet, then descending at an angle singing the Bobolink song. Another was seen swooping down, then flying across a river giving a perfect kingfisher rattle and action.

They very often sing a lovely whisper song that can't be heard more than 5 feet away. When an enemy approaches it gives an angry cat mew, hence the name Catbird. Often it will interrupt its musical phrases with this cat call, a sure identification. Other alarm notes are a harsh chatter, rather wrenlike, also a rattling cry. Do not be surprised to hear a mellow chuck like the soft quack of a duck coming from the center of the Catbird's thicket.

The Mockingbird is widely known for its nocturnal singing, but few know that at times the Catbird also does beautiful nocturnal songs, usually on gorgeous, clear, moonlit nights. One observer described it in her fragrant moonlit garden, "It is then that our cavalier is transformed into a celestial singer, as soul-stirring as the Nightingale in the Old World gardens or the Mockingbird in southern climes."

Last summer the Catbird sang in my yard from May through mid-August, usually perched in the top of the dead apple or the pawpaw tree. Their nest of sticks, weed stems, grass, and leaves was lined with fine leaves and roots. The four eggs of a deeper blue-green than the Robin egg were incubated twelve to thirteen days, and the nestlings remained in the nest ten to twelve days.

The Catbird is seldom parasitized by the Brown-headed Cowbird, but when it is, the Catbird throws a genuine temper tantrum and pitches the hated egg out of the nest.

Carl Kurtz

Mockingbird

The Mockingbird's scientific name, *Mimus polyglottos*, is most appropriate; it means "many-tongued mimic." It was first described by Mark Catesby, early naturalist of the Carolinas, who called it Mock Bird.

A trim gray bird 9 inches long with a long tail, its flashing white wing patches are distinctive in flight and a sure identification mark.

Generally we associate the Mocker with Deep South magnolias and dogwood. We often had the Mocker in our dogwood tree in Georgia and a beautiful sight it was, but not as strikingly beautiful as the gray Mocker perched on a wild Iowa crab apple tree loaded with pink blooms or flitting across purple tops of blooming ironweed as it descends to catch grasshoppers in the bluegrass pasture.

Mockers have been slowly extending their breeding range north for nearly a century. DuMont, in his *Revised List of the Birds of Iowa*, lists thirty-two observations from twenty counties between 1888 and 1933. *Audubon Field Notes* reports observations each year from most of the Canadian provinces and almost every state in the United States.

Mrs. Wayne Collins of Pleasantville, in her "Mockingbirds in Iowa" (*Iowa Bird Life*, March 1967), received from Iowa Ornithologists' Union members records of 69 observations in 1961, 75 in 1962, 80 in 1963, 85 in 1964, and 102 in 1965. Davenport reported 20 to 35 and the Pleasantville area was second with 16 (including young) in 1965. The yearly population has fluctuated, never surpassing 1965's total until this year.

In early May 1970, three were observed along the south side of Red Rock Refuge and one pair in each of three sections in Pleasant Grove Township, near Pleasantville.

Usually, I see the first one in late April. They are summer residents, flying south in late September and October. By nesting time they are settled in old bluegrass pastures with scattered shrubs and cattle-pruned haw trees, the favorite nest site.

As Mrs. Collins wrote, this ideal habitat "had everything desirable to a Mockingbird—insects, nest sites, water, wild fruit of gooseberries, mulberries, red haws, multiflora rose, crab apples, wild cherry, and Virginia creeper." The live-elm singing perches of 1965 are the dead-elm singing perches of 1970. Otherwise, this bird prefers low elevations for feeding and nesting.

Mockers take their domestic duties seriously. Together they build a large nest 2 to 5 feet up in small red haw or Osage orange trees. The outside of the nest is made of thorny sticks and the deep inner cup neatly lined with fine roots. The eggs are blue-green splotched with cinnamon-brown and number from three to six, but four is the usual clutch. Incubation takes about eleven to fourteen days.

The tiny nestlings are dark-skinned with

dark down fore, aft, and on the wings. Most conspicuous are the big mouths with beautiful deep yellow linings.

The nestlings grow rapidly, fed by both parents, remaining in the nest ten to thirteen days, then climbing around the nest tree branches several more days. The young birds' plumage is a little darker gray and brownish on the upper surface, and they are speckle-breasted like most thrush species.

They can do whisper songs in thirty to seventy days, even imitating a frog. The whisper songs are quite lovely and melodious. They also make a scolding note very similar to the adults.

While one Mocker sang to the south, the mother and three young Mockers, each in its own little haw tree, scolded me for half an hour the night of July 27 as I taped on the hillside during a light misty rain.

Birds are not the free creatures we think they are, rather they are "creatures of action and reaction, releases and responses" according to Roger Tory Peterson.

Lynn Phillips owns the big field but the birds couldn't care less. During the nesting season each male defends his territory (sixty-six nests studied in 1965) by song and by action. And the Mocker is one of the most aggressive, flashing across the field to drive away a kingbird or a nest-robbing Blue Jay if they dare approach its nest site.

The adults' ferocity varied greatly when I approached the nest or handled the nestlings for banding. Five pairs scolded—a harsh "chuck" or "whee-e-e"—at a distance, but two adults repeatedly dashed into the tree and at my hand. The sounds on my tape are deafening, screeching cries of deep distress and anger.

A unique behavior trait of the Mocker is its "wing flashing," repeated rapid opening and closing of the wings. I observed this among birds on the ground. Perhaps it is to flush insects from their hiding places in the grass.

Another trait I call the "butterfly flutter" or "flip-flops." Flying straight up about 5 feet above its singing perch, the bird then flutters down to the very same spot, all the while singing.

It is generally conceded that the Mockingbird is the singer supreme on this continent, and many say that its songs surpass those of the European Nightingale and the Skylark. It can mimic the song of any bird it hears, as well as frogs, crickets, and some music.

It has its own beautiful song, too. Two other mimic thrushes sing beautifully but not as well as the Mocker. The Catbird repeats or mimics each phrase once, the Brown Thrasher twice.

On June 16, knowing the nest location of two Mockers, I decided to tape their songs. Little did I realize the frustrations ahead or that the few days planned for the project would stretch to seven weeks of glorious song.

I soon learned that my biggest problem was an overabundance of bird songs—thirty species and those not singing were scolding.

My hours were 7:30 P.M. to 9:30 P.M. daily. The Mocker sings on after sunset when most other species cease singing. I was also quite sure there were three singing Mockers in the big field but could not locate the third nest.

Just west of this field, a Mocker flashed across the road on another evening. This Mocker and its singing tree, a big dead elm, were right beside the road. Here I taped my first song. In one ten-minute segment of tape it changed tunes 117 times. It even did the Whip-poor-will.

It repeated phrases five to seven times and once fifteen times. Its nest and four eggs were in a small Osage orange tree beside the pond in the small pasture.

A few days later I discovered a Mocker to the northeast, its singing tree also a dead elm. I taped forty minutes of its song one evening only to discover the volume was too low. In searching for its nest I located another Mocker to the east with a nest and four young.

Each of the six Mockers I recorded had a favorite singing tree used more often than any other. Their adherence to this tree was especially faithful during nest building, egg laying, and early incubation. I could sit almost directly under their singing trees to record.

I named the high hill where I parked my car Mockingbird Hill because I could stand there and hear Mockers singing from six directions. Their singing seemed inspired during a full moon and they sang a little later too, but every evening was a musical treat.

Fred Kent

Brown Thrasher

If there is one prayer that I treasure above others it is one found in the Book of Common Prayer: "O Heavenly Father, who hast filled the world with beauty; Open, we beseech thee, our eyes to behold thy gracious hand in all thy works." And I do endeavor to get children (and grown-ups, too) to really look and not to pass by unheeding the natural beauty all around.

One evening all five young Urbans—Todd, Janee, Tonja, Christi, and Marci—accompanied me to the Burlington tracks to hunt for Mockingbirds. We were armed only with binoculars and a small mirror on a long collapsible rod for nest snooping.

To our joy we found not just one mimic thrush but all three—the Catbird, Brown Thrasher, and the Mocker—all in the same field. I was careful to impress the children with our good fortune—in very few fields in Iowa could one find all three mimic thrushes. However, the other Mockingbird field also has all three mimics, located a few days earlier.

We found the completed Mocker nest but

no eggs yet. The Brown Thrasher nest was deep in a big brush pile; we did not attempt to reach it with the snooping mirror. We already had found three Brown Thrasher nests in the other Mocker field, all in small cattle-pruned haw trees. We found a Loggerhead Shrike nest with the female incubating six eggs. We viewed them with the mirror in the deep, thick-walled nest in a very thorny Osage orange tree, then we moved quickly on.

The children were finding much of interest in this old bluegrass pasture—plants with pungent aromas, yarrow and two mints, bergamot and catnip. Marci was fascinated by the velvety, feltlike mullein leaves and gathered some for school. There was an old fox den which we decided was not active this spring.

We watched and listened to a Brown Thrasher male singing in a treetop, the first step in the courtship ritual. Later stages often take place on the ground, with the male strutting around the female, his tail spread and drooping on the ground, singing low, often a whisper song. Singing usually ceases after the four to five eggs are laid.

While the Mocker is considered our finest singer, it often mimics some harsh and raucous songs, but I have never heard any harsh songs from the Brown Thrasher. Truly a superb singer, it seems to improvise more than the Mocker, repeating each phrase twice, quite often three times, and occasionally only once like the Catbird.

Incubation period is eleven to twelve days. As a young man Dr. Ira N. Gabrielson (who became a wildlife biologist and director of the U.S. Fish and Wildlife Service, now retired) in 1912 did a detailed study of a brood of young Thrashers in Iowa.

On June 27 from 3:30 A.M. to 9:00 P.M., the male fed the four young 98 times and the female 186 times. The four insects consumed in the largest quantities during the study period were 247 grasshoppers, 425 mayflies, 237 moths, and 103 cutworms.

This was a rare nest on the ground. Usually the nest is 2 to 4 feet above the ground; consequently they are highly vulnerable to predators. I found that 60 to 80 percent of the nests were depredated. This rufous-red bird with white wing bars, a very long tail, a whitish breast streaked with black, and a long, slightly curved bill is the most handsome of our mimic thrushes and the largest, 10 to 11 inches long. Mother called it the Brown Thrush, and I knew it by that name long before I was seven years old.

It nested in tangles of vines and brush in the 2-acre windbreak surrounding the farmhouse. Such windbreaks of deciduous trees are long gone, and the Thrashers are now more often found in old pastures, fencerows, or shrubbery in towns.

Sadly, their habitat is fast disappearing. If suburbanites would plant berry-bearing hedges around their lots, it would provide cover, nest sites, and food for many songbirds.

Permanent residents in Georgia, the Thrashers were a joy on our wooded lot, at our feeders and birdbaths in Warner Robins, Georgia, every day of the year, with a population increase as northern migrants moved in for the winter.

Ground feeders, they simply flip leaves aside with a sideswiping motion of the long bill, then pick up the insects exposed. I also watched them use the bill in a pounding action to dig through 2 inches of rotten wood to get the grubs.

Wild fruit was relished and so were the acorns. They were opened on the ground by a hammering action of the bill, often driving the acorn several inches into the ground before it was hulled and the kernel eaten.

One hot summer day in Georgia, I watched a panting bird leave the nest and four eggs in a tangle of smilax, fly to the feeding area, walk through the birdbath on the ground as if to cool her legs, not stopping to bathe, then walk on to eat cracked corn.

Recently, I observed identical behavior of an incubating bird here, only this birdbath was on a pedestal 10 feet from my window. She simply walked through, then dropped to the ground to feed on cracked corn. Is this a temperature-regulating act? I don't know. However, the Turkey Vulture squirts liquid excreta on its legs, a temperature-regulating mechanism proven by Dr. Daniel Hatch's in-depth study.

Carl Kurtz

Water Pipit

On a crisp, sunshiny November Saturday we had birded the north side of Lake Red Rock and the tailwaters below the dam. We then drove along the west bank and came to a sudden halt.

There, only a few feet away, was a small sparrow-sized bird walking over the bare sandy ground. It was a Water Pipit, a brown, tail-wagging ground bird with a slender bill, a light eye-line, and streakings on the buffy breast and belly which glowed reddish in the light of the setting sun. We watched until it flew in up-and-down flight, revealing the white outer tail feathers. The name comes from its call, "pi-pit"; actually it sounds more like "jee-eet" or "tsip-it."

It is rated uncommon here, usually seen only in October, so this was a late date although we had three on the Red Rock Christmas count on December 29, 1970. They winter across the nation in the southern states from New Jersey and Ohio to the Gulf of Mexico into Guatemala, always in open country, preferably wet.

This small but hardy little bird is circumpolar, nesting in the high Arctic around the top of the world. The grassy nest is placed on the Arctic tundra to northern Alaska, on Mount McKinley above 5,000 feet, across northern Canada, on the moss-covered rocky hills of Labrador, and on the west coast of Greenland.

In the western mountains, it nests in the Alpine zone above timberline. This life zone has been described as a hostile environment, a high cold desert with winds of gale force. During the short summer these Alpine meadows are ablaze with flowers, and their myriad insects furnish an abundant food supply for the pipit nestlings.

This plain little bird courts his mate with a beautiful courtship song flight. An observer on Mount McKinley described it: "The male singing 'che-wee, che-wee' flew straight up 200 feet, then sang steadily as he fluttered his wings and floated down like a falling leaf. The flight was repeated four or five times as the female watched nearby."

Carl Kurtz

Cedar Waxwing

There was a change of pace at my east window feeder one March morning. Instead of Cardinals, nuthatches, and chickadees eating sunflower seeds, thirty-three Cedar Waxwings were eating apples impaled on nails and old home-canned sour cherries (too old for pies, but not spoiled). They also quickly stripped the round bunches of carrion flower (*Smilax*) berries I had gathered in the fall.

They're not a bit fussy about food just so it's fruit. Eighty-seven percent of their diet is fruit and only 13 percent is insects, often hawked in the air. They do have one unusual item in their diet and this is apple blossoms. They actually eat the blossom and prefer yellow transparent blossoms although they have a neighborhood choice of Jonathan, Anoka, Grimes Golden, Jonadel, and Red and Golden Delicious. Eight of fifteen years I watched them eat transparent blossoms in May (this variety bears fruit on alternate

years). One year I tasted all the varieties and could detect no difference in flavor, but the birds knew the difference.

Much has been written about the peaceful, courteous manners of these gregarious birds—half a dozen sitting on a line passing a cherry from beak to beak, etc. With eleven to twenty-one crowded on my feeder I assure you that all was not peace and harmony.

They do have little squabbles. Usually it's just a threatening gesture with crest raised, then lowered, and beak wide open showing the red mouth-lining. Sometimes there is some bill-snapping, and once I saw one peck another's shoulder but no feathers flew. One easily identified by its short outer tail feathers was a real bully.

They thoroughly enjoy the birdbath, with three and four bathing at one time. One morning the bully chased three others out of the bath and then appeared to enjoy a private bath.

To the casual observer, waxwings are all alike—sleek, immaculately groomed, little 5½-inch-long birds. The brown on the crested head and back blends to slate gray in tail and wings, a black mask across the eyes, black bill, dark chin, soft brown breast, yellow belly, and white undertail coverts. A closer look, however, reveals many variations. Most have a quarter-inch border of yellow on the tip of the tail, but a few have only a thin edge of yellow.

The "waxwings," those lovely, red, waxlike appendages or tips attached to the ends of the secondary wing-feathers, vary greatly both in number and length. They were named long ago for their resemblance to red sealing wax, but the modern-day child would call it red plastic. In this flock 66 percent have no red tips at all, but that is no sure criterion of age.

Streaked plumage is the one infallible guide to age, and that is an HY (hatching year) bird.

Robert Yorick, in his thorough review of aging and sexing criteria based on his own data of 258 waxwings banded August-October 1966–1969, at Vischer Ferry, New York, and from his examination of 327 skins at the American Museum of Natural History, found that 51 percent of the young streaked birds in the fall had red tips but 40 percent of his AHY (after hatching year, two years or older) birds had no red tips.

Yorick found other plumage variations. Of his young streaked birds, one had a red tail-band and three had orange tail-bands. Since all four were banded the same day, he raised the question of their origin from a single brood. Only one of my flock has an orange tail-band, and one has no tail-band at all, the entire tail is slate gray. One museum skin was regally attired—eight red tips on the secondaries, three on the primary wing feathers, and eight on the tail, for a total of thirty red wax tips!

Chin-throat color as a sexing criterion, he concluded, also is of low reliability. Velvety black chin-throat color is about 89.5 percent reliable in segregating males and brown-black chin color only 76 percent reliable in segregating females. Ten of my flock have velvety black chins, thirteen have brown-black, and ten have brown chins. Much more banding of streaked juveniles and their recapture in subsequent years is needed to establish reliable criteria.

Why this interest in such criteria? The Banding Laboratory, U.S. Fish and Wildlife Service, sets high standards, demanding that banders record age and sex if reliable criteria are known; if not, the bander is expected to cooperate on studies to establish guidelines.

Little old ladies (like me) in tennis shoes who do not keep abreast of current banding literature and who do not cooperate on research projects are likely to find their banding permits canceled.

The title "cooperator" given banders by the Lab means just that—cooperate or else. Except for State Conservation Commission employees whose job includes banding, banders are unpaid volunteers who must have extensive knowledge of birds. Banding is serious business.

Wandering nomads, waxwings are the most unpredictable of birds. There's no set time for their arrival; it may be any month of the year here. Judging from letters received from all over Iowa, this past winter was a good waxwing year.

In Warner Robins, Georgia, they were a little more predictable, arriving by the thousands shortly before Christmas, followed soon after by Purple Finches (one finch flock turned scavenger, sweeping across our yard daily, cracking and eating the seeds in the waxwing droppings). The waxwings fed on the purple berries of the ligustrum and privet, base plantings at the thousand houses built by one developer. Their favorite resting perches were the utility lines.

The berries were so dry that the waxwings were always thirsty, with flock after flock lining the rims of our three birdbaths, dipping their heads to drink almost with military precision. A pretty sight indeed.

In June 1941, my husband and I were visiting my sister and brother-in-law, the Ray Proffitts, at their farm home near Pleasantville. Hearing a high-pitched "zee, zee"—singers they are not—I located a waxwing nest in the fork of an old box elder tree in the front lawn. Up about 12 feet, a rather bulky nest lined with horsehair, it contained four pretty eggs, blue-gray spotted with purplish brown.

One afternoon the adult birds began a distressed fussing. I investigated, and there was a long snake draped down the tree trunk with its head in the nest. I grabbed the .410 shotgun, loaded it, rushed out, and shot the snake, which fell to the ground. I set up the ladder and began climbing. Halfway up the head end of the snake fell. I dodged just in time to avoid getting hit squarely in the face. The ladder swayed and I grabbed a small branch, thus avoiding a fall. (My language was unprintable!) But I came to the rescue too late, the nest was empty.

Today I'd act with more sense by capturing the snake and releasing it elsewhere. My father appreciated the value of snakes in rodent control. He never allowed one to be killed around the barn, and each hired hand was so instructed.

Only one other waxwing, the circumpolar Bohemian, reaches us and that but rarely. About 7 inches long, it is darker than the Cedar, with rusty undertail coverts and white-and-yellow markings on the wings.

Over the years in Georgia I watched every flock but never saw a Bohemian Waxwing. On December 21, 1957, I saw my first one in Des Moines Waterworks Park while with Ruth Binsfeld's party of observers doing the Christmas Bird Count. This was the first Bohemian Waxwing seen in Des Moines in sixteen years, and not too many have been seen since. I have records in Pleasantville for only two years—two birds in February 1962 and one in December 1966.

The last major invasion in Iowa was reported by the late Arthur J. Palas. On March 18, 1923, he saw a flock of 5,000 feeding on hackberry and wild hemp seeds on Walnut Creek between Des Moines and Valley Junction. In Europe where invasions have been noted for centuries, the largest ever recorded occurred during the autumn and winter of 1965–66.

Larry Stone

Northern Shrike

During the almost unbearable winter of 1977 so hard on humans and birds and beasts, with no northern finches except one lone female Red Crossbill and a few Purple Finches, the reports of forty-one Snowy Owls in twenty-nine Iowa counties were welcome news— pure pleasure to those fortunate observers.

Reports of the Northern Shrike, another rare winter visitor from the far north, was further good news. Dr. Nicholas Halmi of Iowa City, compiler of field reports for *Iowa Bird Life*, official publication of the Iowa Ornithologists' Union, reported eight of these birds in the Iowa City area, ten were seen by Darwin Koenig in northeast Iowa in November, and one was seen by Gene Burns in Dallas County in January. On our Red Rock Christmas Bird Count, December 19, my party made a tentative identification of a

Northern in the south side of the refuge, but the bird flew before we could make a positive identification.

To quote Woodward H. Brown, in his "An Annotated List of the Birds of Iowa," it is a "rare winter resident. There are no recent records from the southern half of the state but on 6 Dec. 1968 one was banded, and on 29 Jan. 1969 another was seen near the state line (by Mrs. Hazel Diggs, Lamoni). Observations have been from 11 Nov. to 22 Feb." My recent records in the Red Rock area included one on November 19, 1967, one on October 27, 1970, and one on October 27, 1973, the last one also seen by Bill Criswell.

The Northern Shrike is a chunky Robin-sized bird 8 to 9 inches long; the Loggerhead is slightly smaller, 7 to 8 inches long. Both are gray above and white to light gray below, with black wings and tail.

Check the following points carefully. The black mask through the eyes ends at the beak in the Northern but is continuous across the top of the beak of the Loggerhead. The base of the lower mandible is light grayish yellow in the Northern; the entire beak is solid black in the Loggerhead. The sides are faintly barred gray in the Northern but plain light grayish white in the Loggerhead. The bill of the Northern is hooked and heavier than the Loggerhead's. The young Northern Shrike is grayish brown with brownish barring on the sides; all young Loggerheads are gray after August.

The Mockingbird is about the same size but has no black mask through the eyes, and the wings and tail are gray, not black.

With the limited statistics available to me, the invasion years of the Snowy Owl and the Northern Shrike appear to be correlated. Both Snowies and Northerns staged major invasions in 1905–06, 1917–18, 1926–27, and 1930–31. The lemmings (two species of arctic mice) are a staple in the diet of both species. During the increase of mice, the predators also increase; when the mice die out the predators exhaust other prey and then either migrate or starve.

The Northern has two methods of hunting, watchful waiting and active pursuit, with the first most often used. By perching quietly on a post, tree, or wire, it is ready to pounce on an unsuspecting mouse. It also hovers over runways of the mice. Other small birds are most often caught by active pursuit. The shrike rises above its victim then dives down, knocking it to the ground with a stunning blow from the powerful beak, which often breaks the small bird's neck. The kill is completed on the ground by a blow at the base of the skull or by biting through the neck.

Having killed its prey, the shrike seizes it with the hooked beak, flies to a nearby thorn tree, and impales the prey on a thorn. With no talons on its feet, the impaling enables the shrike to tear off bite-sized pieces to eat. Although classified a songbird, its digestion is the same as that of the hawks and owls. Powerful juices in the stomach digest everything except bones, fur, and feathers, which are formed into a pellet and ejected or "cast" through the mouth.

This big shrike is circumpolar, found in the eastern hemisphere from Norway to western Siberia, in winter south to central France, northern Italy, western Rumania, and central Russia. In North America, from Alaska and northern Canada to central United States.

It breeds as far north as the spruce forests extend. The thick felted nest is made of coarse grasses, weed stalks, and plant down, reinforced with twigs and profusely lined with white ptarmigan feathers, and placed in spruces or willows 7 to 12 feet up. The five to six eggs are light grayish green marked with blotches of brown according to Bent's *Life Histories.*

Humans were once the shrike's worst enemy. Long ago men were hired to kill the Northern Shrikes on Boston Common, one man killed fifty to protect the newly imported English Sparrow. Today the Northern is welcomed because it reduces the English Sparrow population.

Larry Stone

Loggerhead Shrike

During the past twenty years I have spent many pleasant hours watching the Loggerhead Shrikes, the "butcher birds," on their nesting grounds southeast of Pleasantville. (In the Carolinas the bird often is called the French Mockingbird. Another name in the South is the Cottonpicker because of its bobbing flight over the cotton fields.)

The Mockingbird and the Loggerhead Shrike are both gray birds with white breasts and bellies, about Robin-sized, 8 to 9 inches long. But the Mocker has big white flashing wing patches in flight. The Loggerhead Shrike's wings are blacker and the tail is blacker. The Mocker has an all-gray head and white throat. The Loggerhead Shrike has a gray head with a narrow black facial mask beginning in back of the eye and crossing just above the beak.

The flight of the shrike is very different. On leaving a perch it drops a few feet, then flies with very rapid vibrations of the wings, a fluttering effect in bobbing waves of flight, sweeping upward in a steep glide to the next perch. The flight speed is 22 to 28 miles per hour.

I have been totally fascinated by this species since as a ten-year-old I first saw one impale a freshly killed mouse on a honey locust thorn in our farmyard west of Drakesville in 1919. The "impaling" tree is never the nest tree although it is nearby.

Storage is not the purpose of impaling; the birds do not eat dried or spoiled meat. It is done to hold the prey in position so the shrike can tear off bite-sized pieces with its hooked beak.

This species is an evolutionary "in-between." Classified as a passerine, a perching bird, it often is called the "songbird of prey." It has the hooked bill of a bird of prey, but it has no talons on the feet, and it very rarely carries prey with the feet. It kills mice, frogs, and small birds with a quick cut or snip of the hooked bill on the back of the head or neck, followed by rapid biting motions, sometimes shaking or pounding the prey against the ground or perch.

The summer diet is 72 percent insects, mostly grasshoppers, and 28 percent vertebrates. Birds make up 8 percent of the diet for the year.

A few shrikes overwinter here. On the 1975 Red Rock Christmas Bird Count, Rick and Beth McGeough and Phil Myers counted 3 and I saw 1 west of the count area. Most go farther south. The Christmas counts were 12 in Iowa, 56 in Missouri, 480 in Arkansas, 373 in Louisiana, 175 in Oklahoma, and 2,588 in Texas.

I had one recovery of a banded bird in Arkansas near the Oklahoma line. A word of caution: it requires detailed study to distinguish between the Loggerhead and the Northern Shrike, which rarely reaches us in winter.

The shrikes return here in mid-March, each establishing a territory in its favorite old bluegrass field. Whether it is the male or female that awaits a mate on territory, I do not know.

The courtship is not conspicuous. The male does wing-fluttering and tail-spreading displays, occasionally "tidbitting," offering the female a choice morsel of food. The two have twisting and turning flights across the field.

They are early nesters in April. Of sixty-four nests studied over a ten-year period, all but two were in Osage orange trees. And those two were depredated by house cats, the birds probably were foolish virgins nesting for the first time. Wisely, they placed their second nests in thorn trees.

The nest is a very strong, thick-walled, deep structure of few sticks, mostly weed stalks and fine plant fibers, lined with a few feathers. I usually can recognize the nest at a distance by the gray appearance.

The nest is built long before the Osage orange leafs out, an aid in nest finding. The female builds the nest in ten to twelve days with the male an interested onlooker. The second nest is built in much less time, but not all shrikes nest twice.

I recall one vociferous female that greatly resented my banding her first nestlings only 5 feet up. Her second nest was placed 12 feet

up in a nearby mature Osage orange, secure from the meddling bander. Her first fledglings loafed around this tree during the second incubation, joining their parents in screaming defiance at me every time I approached the tree.

The eggs are light gray with yellowish brown spots. The clutch size usually is given as five to six, rarely seven. However, I had seven egg clutches in eleven of sixty-four nests, a little over 17 percent. The incubation period is longer than that of most songbirds, fifteen to sixteen days.

On hatching, the skin is bright orange with apricot-yellow bill and feet. The skin is smooth, with just a puff of down on elbows and abdomen. The mouth lining is deep pink.

Within an hour of hatching they can raise wobbly heads to beg for food when touched, and they give very thin "tsp, tsp" notes. They are fed very tiny insects at first. Like true birds of prey, they regurgitate indigestible portions of insects or mammals as pellets.

By the time they leave the nest at sixteen days to climb around the nest tree, they are carbon copies of the parent birds. And quite intelligent, they climb higher in the tree at the approach of an enemy. Their location note is "screig."

These birds have suffered a drastic population decline throughout the United States and southern Canada due to persistent pesticides and loss of grasslands, their favorite habitat. The National Audubon Society has placed them on the Blue List of threatened species. The shrike population is thought to be down 90 to 100 percent in some parts of Iowa, and it has been completely extirpated in central Illinois.

Bell's Vireo

John James Audubon on his famed eight-month expedition up the Missouri River discovered the small vireo on May 4, 1843, and named it Bell's Vireo for his companion, J. G. Bell, who collected it and a Harris' Sparrow—two new species—on that day. This was Audubon's last great expedition.

In his notes he says of its haunts: "This species, like other vireos of the smaller class, is usually found in the bottomlands along the shores of the Upper Missouri River, from the neighborhood of the Black Snake Hills as far as we went up that river; finding it in many instances whether in the bottomlands, overgrown with low shrubbery, or along the borders of ravines that discharge the water accumulating during the spring meltings of the snows that cover the upper country prairie land."

Seven members of the vireo family are found in Iowa, the Warbling and the Red-eyed are rated abundant breeding birds; the White-eyed, a southern species, is a rare breeding bird; the Philadelphia and the Solitary are migrants; the Yellow-throated and the Bell's are rated common breeding birds.

Today I would not rate the Bell's a common breeding bird here. Since 1960 there has been a marked population decline. Habitat destruction is one factor, but many undisturbed nesting areas now have no Bell's. Cowbird pressure is heavy, and many areas along highways and railroads have been subject to herbicide "brownouts."

The Bell's Vireo is a tiny bird only 4¼ inches long, head and back gray-olive, wings brownish with thin white wing bars, brown tail, white throat and belly, with a touch of yellow along the sides. The black eyes are encircled with white eye-ring spectacles, and the dark bill is heavy with a tiny hook on the end, both characteristic of vireos. (Sometimes vireos are confused with warblers but their movements are slow and sluggish not fast and darting, and they do not sit in typical upright flycatcher posture.)

Most of the vireos are superb and persistent singers, especially the Red-eyed and the Warbling, but the Bell's song is a series of low phrases, "cheedle, cheedle chee" or "whillowhee, whillowhee, whee," definitely unmusical. The scolding note is a "chee, chee, chee."

The male sings off and on all day, often in the nest shrub, continuing throughout the summer but lessening by August. Musical or not, it is a pleasant sound sung with great enthusiasm, and I regret that I have heard it so seldom in recent years.

All the vireos build tiny exquisite hanging nests suspended from the crotches of thin branches. The Bell's nest is almost always within 3 feet of the ground, rarely as high as 10 feet, with one at 25 feet. I never have found a nest above 4 feet. It is constructed and woven of grayish plant strips, fibers, and leaf fragments and is usually lined with fine grass stems; the rim often appears to be covered with spiderwebs.

Three to five white eggs are sparsely sprinkled with fine brown dots around the large end; occasionally they are spotless. It is during the egg-laying period that the cowbird, a brood parasite, does the greatest damage, tossing out a vireo egg and replacing it with her own. Almost always the Bell's desert this nest and then foolishly build another often within 15 to 25 feet of the old nest.

On a shrubby hillside about Coal Creek I once found three fresh nests close together. Each in turn had been parasitized by the cowbird, and finally the little pair deserted that field. Reports from Texas indicate that the Bell's Vireo sometimes builds a second story above the first nest containing a Dwarf Cowbird egg.

A successful nesting along the Burlington Railroad contained no cowbirds, just four plump Bell's Vireo nestlings, so big that the sides of the nest were stretched and bulging. After banding them, it was difficult to fit them back into the nest. The parent birds came quite close, "chee, cheeing" their anxiety at this outrage.

The diet of the Bell's is almost entirely insects with a surprising number of grasshoppers—very large insects for such a tiny bird. Dr. E. A. Chapin also reported 1.5 percent of the diet to be fruit, but I have never netted one near wild mulberries nor have I seen them eat fruit.

The entire breeding cycle requires more than thirty-five days exclusive of the courtship period—five days for nest building, four or five days for egg laying, thirteen to fourteen days for incubation, and eleven to twelve days as nestlings. However, the fledglings stay

with the parents and are taught to hunt insects for two to four weeks. The earliest spring date recorded in Woodward H. Brown's "Annotated List of the Birds of Iowa" is April 25 with the last fall date September 23.

Jim Messina/Prairie Wings

Yellow Warbler

The wood warblers, guardians of the oaks, have been called the butterflies of the bird world by that eminent ornithologist-painter-author, Roger Tory Peterson.

Thirty-four species have been observed in Iowa, but few remain to nest. The most abundant in Iowa is the Yellowthroat. The Redstart, the southern Yellow-breasted Chat, and the Ovenbird have declined in recent years as more and more woodland is cut and bulldozed.

That most widely distributed member of the warbler family, the Yellow Warbler, ranging from northern Alaska to western Peru and breeding from the Atlantic to the Pacific and from the barren grounds in northern Canada to the Gulf States and Mexico has also declined and is now blue-listed. In the late 1960s, Des Moines had counts of forty-two on their breeding bird census, but now it is a treat to see one.

I must confess that I am a lazy birder, much preferring to sit in a deck chair and observe the behavior of birds and listen to their songs. So it was in the summer of 1976 that I was enjoying the Mockingbirds in Lynn Phillips's old shrubbery, bluegrass cattle pasture.

At times all three mimic thrushes, the Catbird, Brown Thrasher, and Mocker, were singing the evening of June 8 when suddenly there flitted across my line of vision that jewel of a little bird, the male Yellow Warbler, yellow all over with red streaks on the breast and smaller than a Goldfinch. I watched him as he flitted about the top of a wild crab apple tree searching for insects and singing his "see, see, see, ti ti ti, see" constantly.

I began a slow search of the probable nesting territory and saw the female, a paler copy of the male, with a few cinnamon-red streaks along the sides. Now, I was sure it was a breeding pair, but I couldn't locate the nest. Again on June 9, I searched and was scolded by the pair, proof positive of a nesting territory but again no nest found.

On the evening of June 10, I took my chair farther up the hillside and sat partly hidden by low haw trees. Very soon I was rewarded as I saw the little female leave a slender 5-foot haw tree. There, about 4 feet up was the little nest about one-third completed and built of gray plant fibers. Quickly I left the nest territory followed by Lynn's very tame cattle. That evening I heard the Mocker singing the Yellow Warbler song too.

On June 11 the rim of the nest was nearly completed, but a bare branch was still uncovered in the bottom. On June 12 I found a cowbird egg in the nest. The Yellow Warbler builds a second-story nest to cover cowbird eggs, and the egg was already half buried. Now, I don't usually interfere with nature, but the cowbird is probably the Yellow Warbler's greatest enemy, so I removed the cowbird egg. So by my removing the egg, the little bird didn't have to work so hard to build the second story. On June 13 there was no egg in the nest and I was sure they had deserted, although I saw the pair nearby.

The evening of June 14 I was elated to find one small white egg speckled with brown in the neat nest. The clutch of four eggs was completed by June 17, and the female was incubating. The last three eggs had the spot wreathed around the large ends.

From June 17 to June 26 I watched the nest almost daily and always was fearful of predation. The success rate of all species in that field was only 40 to 50 percent in years past.

At 8:00 P.M. on June 26 I found one egg neatly halved and open but the baby bird was not yet out of the shell. On June 27 the nest held three very tiny, very naked pink babies with great bulging dark eyes not open and about four strands of fuzz on the heads. By the evening of June 28 the last egg had hatched and the nest held four babies.

On June 29, in response to my call, Newton wildlife photographer Herb Dorow and wife Edie came and took pictures of the four babies with heads up and tiny beaks wide open begging while the parents fussed and flitted in and out of nearby trees.

By June 30 they had doubled in size. I watched the overworked parents diligently searching for insects in buckbrush, crab apple and haw trees, and tall weeds, feeding the young often from 6:00 A.M. to 9:00 P.M.

By July 4 they were growing rapidly and the nest was crowded. On July 5 I banded the four.

On the evening of July 6 the nest was empty, and I feared predation by snake, mammal, or bird. However, the parents set up a lot of fuss, and finally I glimpsed the olive backs in low shrubs. Bent's *Life Histories* gives the normal time of fledging as nine to twelve days. Mine had fledged on the tenth day.

The breeding cycle had taken a full month, but the parent birds had another two weeks to spend teaching the fledglings to hunt insects before they began their long migration to the tropics. They are early migrants, leaving here from July 15 to July 31.

Carl Kurtz

Yellow-rumped Warbler

On a beautiful October Sunday afternoon my young neighbor, Janee Urban, and I began a field trip to gather data for her science project. Dull and boring research? Indeed not, during fall migration!

We found flocks of migrating land birds, juncos, and White-throated Sparrows with their plaintive, sad calls. A flock of Myrtle Warblers, noted for their swift darting flight, was bathing in the clear water of a small spring-fed brook, a beehive of buzzing, flitting activity. (Officially, they are Yellow-rumped Warblers, the American Ornithologists' Union nomenclature committee has decreed, going back to the original name of long ago. They also have been known as Crowned Wood Warblers and Golden-crowned Flycatchers.) Most warblers are entirely insectivorous, but the Myrtles also like seeds and small fruits, especially wax myrtle berries of the south, hence the name myrtle. In Iowa they eat the white berries of the poison-ivy vine and other small fruits.

During the nesting season they are more widely distributed than most other warblers, from the south slope of the Brooks Range in Alaska across to eastern Labrador, Newfoundland, and Nova Scotia, south into the northern states of the United States.

The loose, bulky nest is saddled on a limb of a conifer from 10 to 50 feet up, always lined with feathers, often of the Ruffed Grouse. The four to five eggs are creamy white with a wreath of cinnamon spots around the large end. Incubation time is twelve to thirteen days, and the nestlings are fed by both parents about two weeks.

The Myrtles are the latest warblers to migrate each fall, and they are the most winter-hardy, often found as far north as southern Illinois and northern Arkansas, though many winter in Central America.

In Georgia the Myrtles ate suet at our feeders all winter. In late March the numbers of males increased, black, white, and gray with a bright yellow rump, cap, and sides. Sometimes fifteen to twenty fluttered like butterflies above the suet at one time. A week or two later the duller-colored females appeared, and we had a repeat performance.

Sid Lipschutz

Pine Warbler

This December 1991 brought back memories of our 1955 December in Georgia. At that time my husband, Wayne, and I had a strange little green bird, not much bigger than a sparrow, come to our suet feeder. It had wing bars. A search of our only bird guide, Peterson's eastern guide, did not help.

My neighbor and friend, Hedvig Cater, couldn't identify it, although she had studied ornithology at the University of Minnesota under Dr. Thomas Roberts, author of the two-volume *Birds of Minnesota*. She, too, had only an eastern guide.

In February, Dr. Dave Johnston of Mercer University in Macon pronounced it a female Western Tanager. Dave took his Ph.D. at the University of California and knew western birds. It was the second Western Tanager ever recorded in Georgia.

After enjoying the bird all winter, I was invited to give my first bird talk to the Georgia Ornithological Society, at the 1956 spring meeting in Rome, Georgia. Little did I realize that it was the first of hundreds of bird programs I would present over the next thirty-five years.

And, on this December 19, 1991, in Iowa, I had a repeat, a tiny green bird on my east feeder. It had a lovely green back with bright yellow throat, breast, and belly, whitish undertail coverts, wing bars, pale streaks along the sides of the breast, dark eyes but no eye rings, and a sharp little beak.

I knew I had a warbler, but which one? I'm no authority on warblers, and I knew it had no business in Iowa in December.

Was it an Orange-crowned, a Nashville, or a Pine? Finally, on January 10, naturalist Tim Schantz photographed it and pronounced it a Pine Warbler, probably a yearling male of such beautiful plumage.

It fed amidst House Finches, juncos, Goldfinches, Cardinals, and chickadees, apparently accustomed to feeding in a mixed flock. When challenged by another bird, it would simply step aside and continue feeding.

I feed only whole oil seeds on that roofed window feeder and it, of course, couldn't manage a whole sunflower seed, but it would pick tiny pieces off the shelf, and it did not spend too long feeding there. So, I mixed peanut butter, oil, and cornmeal and placed a lid of this on the shelf. It wasn't touched for several days. I was so pleased when the bird tasted it, liked it, and ate a few beakfuls. Never greedy, its stay was short, usually between 9:00 and 11:00 A.M. It came in at 3:45 P.M. only once.

Obviously, this bird was feeding elsewhere. Bent, in *Life Histories*, refers to the Pine "Creeper" because it creeps over the tree bark hunting insect eggs. It also eats seed, small fruit, and pine seeds through the winter. In summer, it fly-catches insects on the wing. I placed peanut butter mix on the bark of an old apple tree and watched it do the "creeper" feeding, and once I watched it walking over the ground.

Why should a Pine Warbler choose this neighborhood? I went for a walk and counted all the big pines and spruces on Columbus and Jefferson streets for three blocks south to the wooded ditch and also included acreages west of Jefferson. To my surprise, we have over fifty-six big evergreens and several small ornamentals. Here we had a mixed evergreen and deciduous forest, not too thick, the preferred habitat of the species.

Sad, but this little bird was unaware of our "Alberta Clippers" and subzero weather. Sure enough, it happened. On January 15, the temperature dropped to 7 degrees F below zero, with the windchill factor recorded at 50 degrees F below zero. I woke up in the night and listened to that gale force northwest wind and knew that no warbler could live through that.

And no warbler appeared at the feeder. I was sad and phoned the Bird Hotline to report the loss.

On the morning of January 23, I was standing at my feeder window watching the House Finches and a few Goldfinches when my little Pine "Creeper" flew in and started eating its peanut butter concoction. It looked plump as usual and neatly preened.

It was unbelievable! I didn't believe any warbler could survive that night, so I happily phoned the Bird Hotline and reported the good news.

A. Cruickshank/Vireo

Ovenbird

Did you ever search for a very small bird in a very big timber accompanied by four little sisters aged four to ten—two blondes and two brunettes named Janee, Tonja, Christi, and Marci? Try it sometime, you'll like it!

And if the small bird is an Ovenbird, you can depend on it that you will have more than four companions. No Ovenbird ever established a territory that was not mosquito-infested. Yes indeed, Ovenbirds and mosquitoes go together.

And if you forget the insect repellent and the long-sleeved blouses—which we did on the first field trip—you will have difficulty hearing the loud and ringing "teacher, teacher, teacher" song in a rising crescendo because the swatting of mosquitoes by five pairs of hands pretty well drowns out the "teacher" song.

Carefully I briefed the girls, all daughters of neighbor Greg Urban.

The Ovenbird is one of the many small wood warblers, insect eaters, found only in the New World. Five inches long, it has an orange crown edged with a black line, a white eye ring, olive-brown back and tail, and a white breast streaked with black. The male has the brighter orange crown.

Most warblers feed in the treetops, but the Ovenbird feeds on the ground, walking on the pale pink legs over the leaves or along a fallen log. It is more often heard than seen. It is one of very few ground-nesting warblers and builds a covered, dome-shaped nest of leaves and grasses with a side entrance rather like an old-fashioned dutch oven—hence the name Ovenbird. It lays three to five white eggs spotted with red-brown, and the eggs hatch in twelve days.

Most warblers feed their babies every five to ten minutes, but the Ovenbirds feed theirs every twenty to forty-five minutes. However, they bring enormous meals of grubs, caterpillars, and moths each time. Both sexes feed the young, but the females do most of the feeding. The little Ovenbirds leave the nest long before they are able to fly, wandering over the forest floor calling loud and long for food which the parents continue to bring. Often the male takes part of the brood and the female goes in another direction with the others.

Of course, the girls found much of interest along the very old, narrow, abandoned clay road winding down the bluff and through the woods in Red Rock Refuge. We watched a Cardinal and a Baltimore Oriole. We listened to the "cuc, cuc, cuc" of the Black-billed Cuckoo and the "reep, reep" of a Great Crested Flycatcher.

We were divided in our likes and dislikes of the peculiar perfume of a blooming wild cherry tree. We loved the fading candles on the buckeye tree, the blue phlox, the creamy waxen mayapple blossom shaded by its two umbrella leaves, and the handsome maroon and green calla lily bloom of the jack-in-the-pulpit.

We were delighted to find deer tracks, then the deer crossing, a well-worn path leading up the steep bank and into the woods. Next trip, they asked, could we make plaster casts: my answer, yes.

And finally, we heard the "teacher" song coming from the top of a tall tree. He wasn't singing too much this evening and it was getting late, so we planned another field trip to tape the song.

The second field trip was even more exciting. Armed with insect repellent and long sleeves, we were better able to fend off the clouds of mosquitoes. Only blonde Janee was severely bitten on her shoulders through her sleeves. Prompt medication brought relief.

This evening we heard not just one but four males singing on "territory," the woods ringing with their "teacher, teacher" songs. We searched many little clumps of leaves but not one clump contained an oven-shaped nest! It is one of the most difficult of all nests to find even though the Ovenbird has a small territory of half to one acre.

Using a portable tape recorder, we hoped to tape the songs of all the males for comparison but settled for one. We couldn't be in four places at the same time. Just as we pressed the button to record, we were surprised and pleased to have a flute accompaniment—a Wood Thrush singing its heavenly "ee-o-lay, ee-o-lay" nicely interspersed between the crescendo chants of the Ovenbird song. Truly vesper songs along a cathedral walk!

Yellow-breasted Chat

Would you believe that the warbler family, which is noted for its fine singers, has an eccentric member that exhibits very clownish behavior? It caws, it whistles, it barks, it mews, and from a treetop it flies straight up 8 or 10 feet with wings fluttering, feet dangling, pouring forth musical gurgles and whistles. Then with wings straight up, it hovers for a few seconds before plunging to another nearby perch.

This show-off is expressing a love song, its courtship flight, and is the Yellow-breasted Chat, the largest of the warbler family. Six inches long and Starling-sized but more slender, it has a brownish, olive-green back, wings, and tail, yellow-and-white wing-linings, and a white line extending from the bill and forming an eye ring. Below the white line a spot of black extends between the eye and the bill.

It would appear to be misclassified since it has a large, heavy, and more curved bill, shorter and more rounded wings, and a longer tail than warblers. However, it has nine primary feathers, a partly booted tarsus, and a deeply cleft inner toe; thus, taxonomists have placed it in the warbler family. It differs from the vireo since it has no notch in the bill.

Personally, I would place it with the Mockingbird because of its eccentric behavior and the many mimic notes. However, that authority on bird songs, Aretas Saunders, states that it does not mimic other birds' songs. He also states that "the song of the Chat is not only entirely unlike that of any other warbler, but also unlike that of any other bird with which I am acquainted."

The Chat is a bird I knew well in Georgia and is common throughout the South. I never had seen one in Iowa until some twenty years ago. I was taking a leisurely walk along the Rock Island tracks one May morning enjoying the many bird songs when up popped a Yellow-breasted Chat from a thicket. It lit on the top of a dead tree and proceeded to put on a show. That was a happy summer for me, as I went almost daily through late May, June, and July to watch my clown.

During all the summers since until 1977 I enjoyed the Chats at two different Rock Island crossings a mile apart. Then they failed to return and were on the Blue List, having suffered drastic population declines. Today in 1989 they are no longer on the Blue List, apparently having recovered somewhat.

So it was with delight that I received a phone call from wildlife biologist Robert Thornburg reporting the sight of a Chat along the Rock Island tracks south of his home and west of the two crossings where I had observed them in past years. One beautiful morning in late May 1980 Robert and I drove down the rutty dirt road. We walked and looked and listened. There were bird songs galore but no elusive Chat. As we prepared to leave we saw one peeking out of a dense thicket. A real beauty but no song that day.

A couple of days later I decided to check one of the east crossings hoping a pair might have returned. I walked a considerable distance when I heard a "caw." It was definitely not that of a crow.

He was in a 15-foot elm sapling but flew immediately to the top of a big oak and went into his "flip-up" antics. Twice he did this before hopping from tree to tree, finally disappearing into a thicket.

His voice was distinctive, even when competing with a chorus of Indigo Buntings, Rose-breasted Grosbeaks, Cardinals, Catbirds, Yellowthroats, House Wrens, Great Crested Flycatchers, Field and Song sparrows, orioles, cuckoos, and the distant hooting of a Barred Owl.

In my twenty-five years of bird banding I have banded only two Chats, both in Iowa. On July 10, 1962, a murderously hot day, near the Nelson crossing I banded a Chat that had been captured in a mist net set in a clearing near its favorite singing tree and thicket. On June 1, 1964, I banded one that had been captured in a mist net near the Kimball crossing. Most warblers take the smallest band, size 0, or the next smallest, size 1, but the Chat takes a 1B band, the same size as a bluebird or a Downy Woodpecker.

The nest, placed in the densest thicket, is usually 3 to 5 feet up, and the four to five eggs, white spotted with reddish brown, are incubated about twelve to fourteen days, with both parents feeding the nestlings. The Chat feeds chiefly on insects but does eat some wild fruit such as raspberries, elderberries, and wild grapes.

Dr. Alexander Skutch, longtime resident of Costa Rica, reports that in winter the Chat spreads over Central America and is an abundant resident in the lowlands of Honduras and on both sides of Guatemala, arriving about mid-September and leaving by mid-April.

Burt Shepard

Summer Tanager

On a birdwalk down the old road to the tailwater area of Red Rock Dam in July of 1980, I heard a Robin-like song. It reminded me of a Rose-breasted Grosbeak but was instead a rare Summer Tanager. Its song, though much like the Robin, is more musical than the Rosebreast.

The last time I had heard a Summer Tanager singing was in 1956 in the woods surrounding our Georgia home. It was thrilling, as was the first sighting for all the garden club members. I had, however, seen one in October 1964, when I netted and banded a greenish yellow female or young male at my home in Pleasantville.

Mary Felsing had one, a male, on her birdbath perched beside a male Cardinal two years ago, so there could be no doubt about positive identification.

In May 1974, I saw a male on the suet feeder at Everett and Catherine Griffin's home on Greenwood Drive in Des Moines. Tanagers had nested in that area each summer beginning in 1956, but Jeannette Eyerly in the same neighborhood reports that they have not been seen in the past five years.

Known in Georgia as the Summer Redbird, the tanager nested on a horizontal limb about 15 feet up in our woods. The four eggs were bluish or greenish, liberally sprinkled with brown spots or splotches.

The young male and the female are greenish yellow, but the next spring the male attains his rosy plumage. I once observed one that was a patchwork of red spots on the greenish juvenile plumage, rather like an old-fashioned crazy quilt.

Each evening throughout the summer the rosy red male would fly to the top of our suet feeder tree in Georgia and announce his presence with "pi-tuck, pi-tuck" as he descended the tree. Jeannette reported also that they ate only suet at her feeders over a period of twenty summers. However, I once observed one eating pieces of bread near the kitchen of the Martha Berry School near Rome, Georgia.

The late Woodward H. Brown, in his "Annotated List of the Birds of Iowa," rated the Summer Tanager as "an uncommon breeding bird in the southern half of Iowa" and stated that Anderson in 1907 declared it very rare, with the only described nesting in 1889. DuMont in 1933 mentioned a pair nesting during two or three summers between 1903 and 1906, with no recent records. However, Brown stated that "there has been some northward extension of range in recent years, with observations in Cedar Rapids, Cedar Falls, Wildcat Den, and several nestings in Des Moines from 1956 on." The earliest re-

corded date of arrival is April 20, 1899, in Burlington and the latest October 8, 1966, in Des Moines.

Carl Kurtz

Cardinal

The bird-watching fraternity in the United States has grown by leaps and bounds in recent years. It now numbers well over 5 million.

If you were to ask birders in the eastern part of the country to name their favorite bird at the winter feeding stations, it would undoubtedly be the Cardinal second, with the Robin first choice in the North and the Mockingbird first in the South. It was not always so. The Cardinal is actually a Carolinian or southern species. In Iowa the bird's northward range extension began with the arrival of the first pioneers and did not reach the McGregor area (in northeast Iowa on the Mississippi River) until 1909. Today it is not unusual to have one to two hundred Cardinals on Christmas Bird Counts in southern Ontario, Canada.

I cannot remember my first Redbird, the vernacular name my mother gave it. Certainly, it was before I was four years old. The bright red male with pointed crest, black throat, and orange-red grosbeak and his yellowish brown mate with reddish wings, tail, and crest were at home in the 2-acre windbreak grove surrounding our farm home on the Iowa prairie.

The underbrush provided nest sites in summer. Weed seeds and waste grain provided their winter food. Sometimes during deep snows, bright red Cardinals were scattered among the maroon-red Duroc Jersey hogs in the feedlot.

As a small child my east Tennessee mother told me a story she had heard as a small child from her mother. During the dark days of the Civil War when that area was raided first by the Yankees and then by the Rebels, Grandmother watched a pair of Redbirds build their nest in the old-fashioned yellow rosebush beside the porch. There they raised three baby birds. Two were Cardinals, but one was a "blackbird." An object of great interest, it was, of course, a Brown-headed Cowbird, a brood parasite. However, it was uncommon there.

An enemy of the Cardinal, cowbirds are numerous in Iowa. They will pitch a Cardinal egg to the ground and lay their own egg for the Cardinal to incubate. The Cardinal then assumes the responsibility for feeding the gluttonous nestling cowbird.

One spring day while living in Georgia, I trimmed our broadleaf evergreen hedge, noting an empty Cardinal's nest which I presumed was an old nest. The next morning I discovered my error. The female was on the nest and laid the first egg.

Knowing that I must rectify my mistake or the little bird would surely die sitting in the boiling hot sun for two weeks, I set two tall posts and tacked green awning over a light frame. Much to my relief the birds accepted their new "shade tree." We had the pleasure of watching four little Cardinals grow up and fledge.

The Mockingbird is definitely the dominant yard bird in the South, with all other species giving way. But one day I saw the tables turned. A gray streak flashed past my window hotly pursued by a red streak! It was the Mocker escaping the irate Cardinal father. Thereafter, the Mocker snooped no more near the Cardinal's nest. One morning my next-door neighbor phoned asking me to watch the Cardinal in her pink oleander bush, then covered with fragrant blossoms. I couldn't believe my eyes. This seed-eating bird was snipping off the lovely pink flowers and apparently sipping the nectar in the deep tube. This was a new feeding method to me— one I never saw repeated.

My husband and I watched the courtship of this pair from the first "what cheer, what cheer," sung by the male in January, then his solicitous "tidbitting" the female with special goodies. This he continued during the incubation period, although she often left the nest for short feeding forays.

Both devoted parents fed the nestlings insects at least fifteen hours every day. After fledging, the four continued begging and crying for food until their beaks had hardened enough to do their own seed hulling. All winter long the family remained with us, eating seeds at our Saint Francis shrine box-feeder and cracking the seed pods of our mimosa tree, neatly extracting the seeds. The Cardinal is an early bird, and many a morning we were awakened by the noisy crackling sound just outside our bedroom.

Albinism, the absence of color, is fairly common among House Sparrows but rather uncommon among Cardinals. We once watched a female albino Cardinal which was mated with a bright red male. They reared two normal-colored little Cardinals. Albinism is carried by a recessive gene. However, if both parents had been pure albinos, snow-white with pink eyes, all the nestlings would have been white.

Often an albino is shunned and driven away from food. This strikingly beautiful pair of Cardinals was an exception to the rule.

Carl Kurtz

Rose-breasted Grosbeak

As Maxine, Sheri, and I slowly walked the shrubby old pasture one late June morning enjoying the many birds and hunting without success for the tiny hanging nest of the sing-ing Bell's Vireo, we came upon a concentra-tion of many species flying in and out to feed on the ripe fruit of a mulberry tree. And there we watched two singing male Rose-breasted Grosbeaks and one female. I was unaware of this species in the old field, but I had not checked the tangle of shrubs in the southwest corner of the field and they flew in that direction.

The handsome male is black and white on the back, a rose triangular patch on the throat and breast, a white belly, and a heavy white

beak. The female in protective coloration of buffy brown streaks and grayish white has beautiful golden yellow wing-linings.

Schoolchildren, also adults, always gasp with delight when I am banding at Outdoor Education Days and I open the wing to show them the rose lining of the male and the gold lining of the female. I also hold them very, very carefully to avoid being bitten. The heavy beak can shuck a sunflower seed in two seconds flat and raise a black-and-blue welt on a finger in less time than that.

Unlike the neat substantial nest of its cousin the Cardinal, the Rosebreast builds a thin, flimsy nest to hold its three to five greenish blue eggs, spotted with brown. The male is a devoted spouse, sharing the incuba-tion duties with his plain mate for twelve to fourteen days. Several years ago I found five Rosebreast nests in shrubs and saplings along the Rock Island Railroad, and on two of the nests the males were incubating and singing their soft Robin-like songs.

Very rarely they sing at night. Usually the song season ends by mid-July, although a young male may sing in late August or Sep-tember. The call note is a sharp metallic "click."

Both parents feed the nestlings a diet of in-sects and larvae with a few seeds and fruit. The adults are seedeaters, often coming to feeders, but they also relish many wild fruits.

They winter in Central America, and un-fortunately many are caught and kept as cage birds (also Indigo and Painted buntings). It is illegal to cage any wild bird in the United States and Canada.

The extreme longevity in the wild is eleven years according to banding records, but they have been known to live fifteen to twenty-four years when caged under ideal conditions.

In one caged bird that lived to be fifteen years old, the white of the plumage finally became a very beautiful rosy red.

Carl Kurtz

Indigo Bunting

The panoramic view from the bluff at Southview Marina on the north shore of Lake Red Rock is magnificent—sailboats, motorboats, and launches on the lake, the wooded bluffs on both sides of the lake, and, of course, the many songbirds during the nesting season singing their territorial songs. If you would like to hear the call of the Whip-poor-will, this is the spot to come at dusk. The repetitious "whip-poor-will" call goes on for hours each night, with several males calling. He is not a great musician, but the call is pleasant, often heard along with the eight deep hoots of the Barred Owl.

The incessant daytime singers during a July picnic supper were the beautiful little Indigo Bunting males, one to the north in a treetop and one below us in a treetop close to the lakeshore. A singing Song Sparrow lacked the persistence of the Indigo, and the "reep, reep" of the Great Crested Flycatcher was but a rude interruption to the "sweet, sweet, where, where, here, here" Indigo song. An overflight by that Beau Brummell of the duck world, a male Wood Duck, was completely ignored.

The Indigo sings three types of songs: the advertising song, which we were hearing, always sung from a high perch; a squeaky song given in territorial defense when an enemy is too near the nest; and the flight song. No songbird sings longer each day than the Indigo—twelve hours, with the other time spent in rest and feeding.

Few birds sing during the midday heat, but the Indigo is not deterred by high temperatures. The hotter the day, the longer he sings. T. S. Roberts estimated that at the rate of 6 songs a minute, the Indigo sings 4,320 songs a day. And they sing from May through mid-August, much later in the season than other songbirds.

The Indigo is a breeding bird through the eastern United States and southern Canada, and the Lazuli Bunting is the western counterpart with some overlapping and hybridization. In a recent study Emlen, Rising, and Thompson reported, "The ranges of the Indigo and Lazuli Buntings in the plains seem to be in a dynamic state. In Nebraska the Indigo Bunting has apparently displaced the Lazuli Bunting westward, perhaps as much as 215 kilometers (roughly 135 miles) in 15 years, 1955 to 1969."

A small bird, only 4½ inches long, about the same size as the Goldfinch, a seedeater with a stout, conical black bill, the male is a deep indigo-blue, almost iridescent, with brownish black wings and tail. The female and young birds are plain brown with no markings except a tinge of blue on the tail or shoulders. The adult male acquires a brownish plumage in fall before departure for the winter grounds in Central America.

They stay in Iowa from early May to mid-October, although my own late date is September 25. In Georgia they arrived about April 26, eating cracked corn and aggressively driving sparrows away. I have never had them at my Iowa feeders, however.

The nest is neat, usually low and rarely as high as 15 feet, built by the female, a skulker preferring to stay hidden low in thickets. The four to five bluish white eggs are incubated by the female twelve to thirteen days. Nestlings are fed insects by both parents and fledge at ten to thirteen days. Both sexes are brown until the second year.

One mid-June Todd Urban and I did a breeding bird census along the Rock Island Railroad. In a half-mile stretch we found six singing males in ideal habitat—many thickets for the low nests and trees and telegraph wires for the high singing-perches of the males. These two requirements are a must for Indigo breeding grounds. You will see the male a hundred times to one glimpse of the shy little brown female.

Carl Kurtz

Tree Sparrow

Of all our winter finches and sparrows, the juncos and Tree Sparrows are the most numerous and the most dependable. They are not, however, as numerous as they once were.

On a beautiful day of sunshine, blue sky, and puffy white clouds in mid-January of 1980, Gene and Maxine Crane and I set out in Gene's four-wheel-drive pickup to do a count of sparrows in Red Rock Refuge, north of Pleasantville. As usual, we found no birds along the roadsides until we hit the bumpy, rutty, abandoned roads in the floodplain. Near the old Hunt farm we counted fifty juncos, the only ones we saw all morning. There is no doubt their population has declined during the last twenty years.

Turning west, we bumped over a mile of very rough road and observed several flocks of Tree Sparrows numbering in the hundreds. Perky little birds, fairly fearless, they kept popping up on tall weed stems to take a look at us, then dropping down to continue feeding on weed seeds. Their rusty red heads glowed in the bright sunlight, and the dark "stickpin" spots on their clear gray breasts were conspicuous.

We continued our bumpy ride all the way across the floodplain to the site of the old Bennington Bridge that was torn down several years ago. The Des Moines River had a few spots of open water but no waterfowl.

Returning, we stopped at an old gravel pit and watched another large flock of Tree Sparrows flitting about on the ice and drinking from tiny open water holes near the bank.

All morning a Marsh Hawk hunted back and forth, flying low in zigzags across the vast fields looking for birds or mice, but we never saw it make a kill. A beautiful, graceful bird, it is our only harrier.

Altogether, we saw at least a thousand Tree Sparrows. All were in friendly flocks with no single birds. Programmed for cool weather, winter and summer, they also sing all winter. Thoreau once described the melodious warbling twitter of a winter flock as like the "tinkle of icicles."

Some forty years ago, Marguerite Baumgartner, beginning her graduate study on the life history of the Tree Sparrow, wrote hopefully to Arthur Cleveland Bent (author of *Life Histories*) requesting information. He replied that such information was sparse and stated, "It is regrettable that you did not select a bird you could study in the field yourself." Challenged by that comment, Marguerite spent three months on the nesting grounds—June through August in 1934—near Churchill, Manitoba.

She found that song, not fighting, was the chief territorial defense, and although the birds were numerous, the nests on the ground in the scrub birch-willow thickets were difficult to find. However, twenty-six nests were studied in depth. The birds arrived about May 25 and they left by September 29. The first few days are devoted to courtship, then June 5 through June 13 to nest building. The four to six eggs, dotted and scrawled with brown on a pale blue or greenish background, are laid during the period June 11–28.

Incubation by the female requires twelve to thirteen days. Both parents feed the nestlings for nine to ten days, and they continue to feed the fledglings for about two weeks after they leave the nest. Fall flocking begins about August 17–19, and all have departed by September 29.

In 1897, a food analyst found that the Tree Sparrow eats about one-quarter of an ounce of seeds per day. At that time, he estimated that ten birds per square mile spend an average of 200 days in Iowa, destroying 875 tons of weed and grass seeds annually. It was not unusual to find 700 to 982 seeds in the crop alone, with another 200 in a crushed mass in the stomach.

The center of abundance of these sparrows in a normal winter is through central Nebraska, Kansas, Missouri, and Iowa and across the corn states to the east, according to Christmas counts and banding returns. During a mild winter, many stay in the northern states and in Canada. A severe winter may drive them farther south to Oklahoma. Altogether, they journey some 1,500 to 2,000 miles from nesting to winter grounds. Alaska birds travel some 3,000 miles. Believe it or not, their migration is segregated, with the females wintering farther south than the males.

Carl Kurtz

Chipping Sparrow

All my life I have been fascinated and thrilled by the nesting of birds, the songs, the courtship rituals, the nest building—many are superb architects—the beautiful eggs, the tiny naked nestlings of songbirds, the devoted parents carrying food all the long day through, often fifteen hours with only brief rests.

No need to wander far afield to do nest studies; many species prefer nest sites in town. Every year the dependable little Chipping Sparrow nests here singing over and over its sometimes monotonous but always cheery "chip, chip, chip" song, warning other Chippies to stay away from its nesting territory.

Of the nineteen sparrow species found in Iowa, only six are common breeding birds with five preferring nest sites in the country. Our little Chippie with the rusty cap and black-and-white eye lines likes small evergreen trees, but almost any shrub will do.

A couple of years ago a pair nested in the Lester Graves's old-fashioned, heavily perfumed, white rosebush only 2½ feet above the ground and 4 feet from a window. Last year a pair nested in a privet shrub beside the door of the Dean Graves's home, and my pair nested in the top of the old Concord grape arbor.

I have never found a nest higher than 6 feet, but W. D. Stull at Lake Itasca, Minnesota, found 39 percent of the nests above 6 six feet, and two studied in Itasca State Park by R. Galati were located at 56½ feet and 54½ feet in black spruce trees, each within 2 feet of the treetop.

The small nest is a work of art constructed by the female of dead grass, weed stems, and rootlets and lined with horsehair. If none is available, fine rootlets and hair of cattle, deer, or other animals are used. Stull carried horsehair with him to Itasca, using it as bait to lure the females into traps for banding during his study.

Of all birds' eggs, the Chippie's small blue egg wreathed around the large end with brown spots is the loveliest to me. Color and spots, however, serve a practical purpose, that of camouflage in open nests. Incubation of the four eggs requires from eleven to fourteen days depending on the air temperature. At Itasca, Stull found that a temperature of 48 degrees F lengthened the incubation to fourteen days.

Both parents feed the nestlings on insects. One observer reported the parents fed their seven-day-old nestlings 189 times in a long day of fifteen hours and forty-five minutes, from 4:05 A.M. to 7:50 P.M.

Growth is rapid and banding must be done by the sixth day. If it's done later, they will jump from the nest and be lost. Many are able to leave the nest on the ninth day, hopping about the tree and flying well by the fourteenth day. Fledglings have light, streaked breasts and heads and no rusty cap.

Chief enemies are the many cats in town, with the predation rate as high as 80 percent, much higher than in the country although no field is free of marauding cats. (The big Mockingbird field is the hunting ground for cats from three nearby farmhouses, and no Mockers nested successfully there last year— the toll of songbird nestlings killed by cats each year is enormous.)

Cowbirds are reported in the literature to be a negligible factor in town, but I have found their eggs in Chipping Sparrow, Indigo Bunting, and Yellowthroat nests in Pleasantville.

My villainous little wrens nesting in the box attached to my window frame for easy viewing are ruthless predators, puncturing the eggs of any small birds nesting within their territory, thus conserving the insect supply for their six to eight nestlings. Each time I find hypodermic punctures in Chippie or Indigo Bunting eggs, I swear I'll burn that wren box, but I always relent.

The Chippies' diet is roughly 62 percent weed and grass seeds and 38 percent insects, but during the summer months it runs to 93 percent insects. They ate cracked corn throughout the winter at our Georgia feeders, and on warm days they caught insects in the air, flycatcher fashion. They also engaged in aerial combat, "rooster-type" fights in the air.

Footpox, a viral disease, was rampant in the East during the 1950s, and some of the Chippies at our Georgia feeders had tumors on toes or the end of a toe or two missing, but they were not as seriously affected as the towhees. I have never seen a case of footpox in Iowa.

One of the most widely distributed of the sparrows, the Chippie nests in most of the states from coast to coast, into southern Canada and far into the South, wintering in the southern states as far north as southern Arkansas. Woodward H. Brown, in "An An-notated List of the Birds of Iowa," rates it a common breeding bird, with the earliest spring date March 28 and fall migration ending by late October but with occasional stragglers through the end of the year.

Larry Stone

Grasshopper Sparrow

This shy, secretive little sparrow so often unnoticed because of its insectlike, buzzy "grasshopper" song and its retiring ways is strictly a grassland bird. The Grasshopper Sparrow prefers cultivated grasslands, and its favorite is now brome and orchard grass, although I have found it in red-clover fields, seldom in bluegrass unless it was unpastured, and never in oat fields.

It will desert a field when it becomes a third overgrown with brush. In recent years many hay fields and pastures have been converted to row crops and roadsides too often saturated with herbicides. Consequently, this species has suffered a dramatic population decline and is now blue-listed. Watch for it.

Short-tailed and flat-headed, with a pale stripe through the center of the crown, it has a heavily striped brown-and-tan back and an unstreaked, buffy breast. It is smaller than the English Sparrow. For many years I could be sure of hearing the "grasshopper" song in brome-grass pasture along the south side of Red Rock Refuge. The Bobolink's "spink, spank, spink," the Redwing's "okalee," and the Dickcissel, Brown Thrasher, and Yellow Warbler at times drown out the insect buzz of the Grasshopper Sparrow on its fencepost singing perch.

From his singing perch, the male Grasshopper Sparrow declared his territory of perhaps 1½ acres. The latter part of April is arrival time; the early date for Iowa is April 11, and I never see them before May 1.

The male arrives first and establishes his territory. His buzzy song is a warning to the other males to keep out. No great fighter, he displays his hostility by crouching and rapid wing-fluttering, then standing tall takes his turn "grasshopper" singing when his neighbor ceases. This song duel is often heard throughout the day. Occasionally, if the aerial territory is invaded, the invader is chased out, but generally the singing suffices.

Bent, in *Life Histories*, reports that "night singing, particularly when the moon is full, is a common habit with both eastern and western races of this species." During the last of May and June while the parents are busy feeding the young, the song is heard less often.

The ground nest, built of stems and blades of grass lined with fine grass-rootlets and sometimes horsehair, is usually domed at the back, giving it an ovenlike appearance. First nests are usually built in mid- or late May and the second nest in late June or July. The four or five creamy-white eggs speckled with reddish brown are incubated eleven or twelve days by the female. On leaving the nest, she runs a few feet through the grass before fly-ing. Returning, she never flies directly to the nest but drops into the grass several feet away. If discovered, she leads the intruder away from the nest with the crippled wing act. The nestlings are fed small insects and by the ninth day are able to leave the nest and run mouselike through the grass.

Grasshopper Sparrows are chiefly insectivorous, with grasshoppers the favorite. Caterpillars, cutworms, armyworms, beetles, leafhoppers, and many other insects are taken. Weed seeds make up 37 percent of the total diet.

Found in suitable habitat throughout the United States and the southern edge of Canada in summer, they winter in the southern states, the West Indies, Mexico, and Guatemala. In checking through the National Audubon Society's *American Birds* 1974 Christmas Bird Counts, I found only one in Louisiana, one in Alabama, one in South Carolina, and seven in Georgia. Texas had fifty-three and Florida forty-three, but many of those are permanent residents. There were three on the Belize count and seven in Puerto Rico.

Carl Kurtz

Harris' Sparrow

The Harris' Sparrow, largest of the nineteen sparrows reaching Iowa, is distributed in a narrow band from north to south down the very center of the continent. The principal breeding grounds are in the strip of stunted timber extending 800 miles between Hudson Bay and Great Bear Lake, along the northern border of the transcontinental forest.

The normal winter range is a narrow strip some 200 to 900 miles wide from southeastern Nebraska (with 434 on its 1974 Christmas Bird Count) to Kansas (10,227 in the 1974 count), and into Oklahoma to central Texas. Small flocks are found on either side of this strip, with Sioux City reporting from 9 to 69 the past ten years. Shenandoah in southwest Iowa has the high count each year, from 30 to 163. The 1974 total for Iowa was 261. Red Rock had only 4, and our 1970 count of 42 was an all-time high and most unusual.

Thomas Nuttall, the ornithologist, was the discoverer of this species, collecting the first one April 28, 1834, on an expedition up the Missouri River. On May 13, 1834, Maximilian, Prince of Wied, was returning from a similar expedition when he also collected one. Neither man published his discoveries for several years, so in 1843 when Audubon's companion, Edward Harris, collected this species near Fort Leavenworth, Kansas, while the two were on a steamboat trip up the Missouri, Audubon named the bird for his companion, Harris, not knowing that Nuttall already had given it a more appropriate name, the Mourning Finch.

It was almost one hundred years after its discovery in 1834 before George Miksch Sutton of the Carnegie Museum–Cornell University team located a nest and eggs in the Churchill, Manitoba, area in mid-June 1931 at the edge of a clump of fairly tall spruce trees on top of a water-bound, mossy hummock about 13 inches above the brown water. This distinguished man of science and art described his personal feelings: "As I knelt to examine the nest a thrill the like of which I had never felt before passed through me. And I talked aloud! 'Here!' I said, 'here in this beautiful place!' At my fingertips lay treasures that were beyond price. Mine was man's first glimpse of the eggs of the Harris' Sparrow, in the lovely bird's wilderness home."

That nest and nine others found in the next three weeks, all on mossy hummocks, chiefly were built of grass, no hair, feathers, or plant down. The one nest was located about 2 miles back from the Barren Grounds and contained from three to five white or greenish white eggs heavily speckled, spotted, or splotched with brown.

The birds are extremely shy and wary on the breeding grounds, and photographing them near the nest is almost impossible. The minimum incubation period of thirteen and a half days was not determined until 1966 by Jehl and Hussell. The food of the nestlings has not yet been studied in depth but is presumed to be the abundant insects of that area.

The handsome sparrow is the size of a

Rose-breasted Grosbeak, 7 to 7¾ inches long but more slender. The full breeding plumage is recognized easily—the solid black crown and throat, the pink conical bill, the streaked sides, the white belly, and the brown-streaked back—but the intergradations range from an all-white throat and tannish crown through a variety of splotched and speckled throats.

Their fall migration is a leisurely affair, often covering three months as they eat weed seeds along the way. In October and November 1964, a flock of twenty lingered here more than a month, feeding in my neighbor's acreage and along the weedy, shrubby ditches to the southwest. I banded eleven of the flock and two repeated, one a month after banding. Unlike their behavior on the breeding grounds, they are unwary and unsuspicious during migration. All that I have handled have been quite gentle, never biting or struggling. Other banders report frequent repeaters in the traps.

Woodward H. Brown, in his 1971 "Annotated List of the Birds of Iowa," rates this species as "a common migrant in the west, uncommon in the east and a rare winter resident, with fall migrants seen from early October to mid-November and those in spring from mid-March to early May. It is found usually in small numbers in widely scattered locations in winter."

I had three through January, February, and March 1968 at my feeders and one through January and February 1969. My earliest fall date was October 8, 1964, and my latest spring date was May 12, 1970, in Red Rock Refuge. Brown records September 23 and May 26 as extreme dates.

According to Baumgartner's eighteen-year study in Oklahoma where she banded hundreds, their winter territory is fairly small and their attachment to this territory is strong, returning to it many winters. Their homing instinct is excellent; when released 2 miles from their winter territory, most homed successfully. Her longevity study of 121 birds showed a loss of 57 in the first two years, with only 7 surviving to age seven. The Banding Office, with thousands of Harris' Sparrow records during a period of twenty-five years, has at least one record of an eight-and-a-half-year-old bird, the longevity record to date.

Their food habits are beyond reproach: 92 percent weed and grass seeds, sunflower seeds, and waste corn and 8 percent insects. At feeders they are attracted by mixed bird seed, sunflower seeds, cracked corn, hemp, millet, grain, and sorghum—rarely crumbs and suet. They also feed on poison-ivy berries and occasionally in spring on the buds and blossoms of Chinese elm. However, they are primarily ground feeders, kicking and scratching among the fallen leaves and weeds searching for seeds and insects. Generally, they are found near underbrush; I find them mostly in the weed fields along the south edge of Red Rock Refuge. When disturbed, they fly up into trees and shrubs unlike most of the sparrows that go lower and out of sight.

Through the winter at my home and on spring and fall field trips I have heard them occasionally give a plaintive whistle, a little like that of the White-throated Sparrow I heard so often during Georgia winters. (The Harris' never reached us in Georgia.) The call note is "tchip" or "cheenk."

During the spring migration the singing is more prolonged, often a chorus. Margaret M. Nice described the song as being of exquisite sweetness, the very spirit of serenity and peace. I agree. They sing constantly on the breeding grounds.

Dark-eyed Junco

Feeder trouble is not uncommon, as anyone who maintains a winter feeding station well knows—squirrels, sparrows, Starlings, cats—you name it, I've had it. But this winter it's been double trouble with the "double-scratch" of the Dark-eyed Junco, formerly the Slate-colored Junco, so aptly described as "leaden skies above, snow below" and often called the Snow Bird.

And what, you might ask, could this slender little sparrow-sized bird do that was so exasperating? Certain of our New World sparrows—including the Tree, Song, Swamp, and White-throated sparrows—the juncos, and the towhees very effectively forage for food on the ground, using the "double-scratch." These sparrows all hop and seldom walk. They scratch by hopping forward and then back with both feet at once. Actually, it is a double kick.

Now, imagine this small bird doing the "double-scratch" in the rusty tuna can of sunflower seeds on my east window shelf. I assure you, the explosion of seeds pinging against the glass sounds like buckshot. The first time it happened I prepared to hit the floor, when I glanced at the mirror on the west wall and realized that a little junco had scared me stiff.

In no time at all he had very nearly emptied the can in his search for a seed slender enough to hull with his small pink bill. Given a chance at the feeders, the junco diet will be about 80 percent wild seeds, less than 20 percent cracked corn, and only .5 or 1 percent sunflower seeds. Hard to believe, as they constantly emptied the can. It required one to six minutes for a junco to hull one sunflower seed, whereas a Cardinal can hull one in a few seconds.

I tried a half-inch mesh cover cut from a potato bag but that was too stiff. It wouldn't sink down on the seeds, and only the long-billed Blue Jays could reach them. Finally, by imposing on neighbor George Dyer, the problem was solved. He owns a beautiful antique coffee grinder, the big kind with two 15-inch iron wheels used in grocery stores half a century ago but now powered by an electric motor. George set it for coarse grind and cracked the sunflower seeds. Success at last. Now I'm much happier and so are the juncos. They eat contentedly and seldom scratch.

The juncos are fierce little fighters around any feeding area, especially if several are present. This may be amusing to watch, but it is an important survival factor—survival of the fittest.

Researcher Steve Fretwell and others have pretty well verified the dominance—food supply—survival hypothesis for junco flocks. The weed patches available are taken over by the dominant group. Perhaps 50 percent of the total flock have feeding positions, and their survival rate is roughly 89 percent. The intermediate dominant group has a 40 percent survival rate, and the subdominant group has a survival rate of close to zero. Speaking plainly, they starve to death.

It is probably the members of this subdominant group that we see at our feeding stations. The juncos I banded near lush weed fields did not appear at my feeders, though I did a count occasionally as they came in to roost in the cemetery evergreens.

However, over a sixteen-year period, the junco population has declined as the use of herbicides has increased. It is only in the Red Rock Refuge where spraying of all insecticides and herbicides is banned that the population remains fairly high. Three junco species are found in North America, but the Dark-eyed is the only one found throughout, breeding from the tree limits of Canada and Alaska southward into the northern states and wintering all over the United States except the desert Southwest and southern Florida. The Oregon Junco, a western form, reaches us rarely. I had one this winter and Nola Vander Streek had one in Pella.

The Carolina Dark-eyed Junco, the race found in the Appalachian Mountains, is almost the only bird in the eastern United States to practice "vertical migration," where in a vertical mountain mile the life-zones change as much as in a 2,500-mile level flight from the southern states to northern Canada. In fall they descend the mountain and in spring ascend to nest.

Usually they nest on the ground, but Alexander Sprunt II had them nesting on rafters in his garage and one in a swinging fern basket on the front porch of his North Caro-lina mountain home. The four to six bluish white eggs speckled with brown are incubated twelve to thirteen days, and the nestlings, fed entirely on insects, are fledged at nine to twelve days.

Once long ago while walking on the Appalachian Trail in North Carolina and duly admiring the flora—a beautiful cucumber tree, patches of glassy green galax—I came to an opening and there was a little family flock of juncos busily working over a weed patch, a delightful scene in a magnificent setting.

The song is a simple lovely trill, and sometimes I have heard it here in February or March long before the pair bond was established. Louise de Kiriline Lawrence of Canada, well known for her scientific accuracy and the poetic quality of her writings, describes "the lovely, tinkling chorus by the juncos in early spring," as if a myriad of woodland sprites were shaking little bells in an intensive competition—"Tililililili, tililili-tililli."

Lapland Longspur

We were in western Dallas County early in March on our way to Springbrook State Park near Guthrie Center when we began to see huge flocks of small birds swirling over the level, barren fields, sometimes touching down and then flying up and away again. From the peculiar, clipped-wing flight pattern, we easily identified many, but not all, as Horned Larks. Snow Buntings often associate with Horned Larks, but these birds appeared to be too dark to be buntings. Finally, we came upon a flock running about on the shoulder of the road and were able to identify many as Lapland Longspurs.

The size of sparrows, they also resemble sparrows in color but have an extralong, spur-like nail on the hind toe. The back of the neck is chestnut-colored, the throat is white with a half moon of gray-black across the upper chest. This is their winter plumage.

The female is plain, streaked with no distinctive pattern, but we saw no females. The males migrate ahead of the females.

Intermittent snow flurries bothered these little cold-weather birds not at all. They were beginning the long journey to their breeding grounds on the tundra. It would be at least two and a half months before their arrival in the far north.

It is a long, hazardous, and sometimes disastrous journey. I reread the account by Dr. Thomas Roberts in *Birds of Minnesota* (1932) of a catastrophe that occurred March 13–14, 1904, during the peak of the longspur migration. A wet, heavy snow fell on a dark night in southwestern Minnesota and northwestern Iowa. Roberts estimated that millions of longspurs died in this 1,500–square mile area. On the ice of two small lakes, an area of about 2 square miles, at least 750,000 longspurs lay dead the morning of March 14.

By the time the males arrive on the tundra in late May they will be dressed in handsome nuptial plumage—the top of the head, the throat, and the chest are a deep jet-black with a buffy stripe running back from the eye and forming a collar on the back of the neck.

The extensive circumpolar breeding range of this species is equaled by few other birds. It extends from the tundra of Kamchatka west through Siberia to Scandinavia, Franz Josef Land, Greenland, northern Canada, and Alaska. The generally high density of nesting birds is likely unequaled by other birds, according to longspur authority Francis S. Williamson. Densities range from 568 birds per square mile on Prince of Wales Island to one pair per 5 to 15 acres on Baffin Island. Even greater densities have been recorded in Alaska.

According to one account, a few males arrive on Southampton Island between May 20 and 29, with large flocks of males arriving June 1 and the females a few days later. On June 8, the observer heard the first flight song. When one male began to sing, they all sang. The females laid their two to six brown-speckled pale green eggs at about the same time in nests on wet meadows. The young remain in the nest only eight or nine days. The short Arctic summer demands fast action for survival of the species.

Jim Messina/Prairie Wings

Bobolink

William Cullen Bryant's poem, "Robert of Lincoln," with its "bobolink, bobolink spink, spank, spink," was a favorite of mine as a schoolchild. No longer fashionable in educational circles, Bryant is now excluded from most elementary school readers. Too deep, they say.

Bobolink, the common name derived from the song, was once one of many vernacular names given this bird by the market gunners. (Protective laws passed around 1903 probably saved the Bobolink from extinction. A gourmet dish, in one season 720,000 Bobolinks were shipped from just one small South Carolina town.)

Other vernacular names include Skunk bird from a Cree Indian term referring to its black-and-white striped back, Goglu as its song sounds to the French in Quebec, and Rice bird in the southeastern states for its depredation of the rice fields.

Of all small land birds, the Bobolink is the top migrant, making a round-trip each year of roughly 14,000 miles from central Canada to the Argentine pampas, island-hopping through the Caribbean. Originally an eastern bird, the Bobolink moved westward with the pioneers' hay and cereal crops, reaching Oregon by the turn of the century. But it still retains the old migration routes, flying east and southeast to the Atlantic and then south to Brazil and Argentina instead of taking the much shorter Mexican route.

I knew the bird as a child during two summers in Wisconsin but not in Iowa. I observed many in Georgia dressed in their sparrow-colored winter coats (acquired during the late July molt) at the Savannah National Wildlife Refuge in September and October and again in bright nuptial plumage as a spring migrant. It's rated a common breeding bird in Iowa, but I never find it in more than two or three fields each summer. In hay fields, the early June mowing destroys the nesting; in permanent pasture it is more successful.

The Bobolinks arrive in early May at their favorite pasture on the Paul Harp farm bounded on the north by Red Rock Refuge, with Red Rock Bluff a short way east. The gaudily attired males arrive about a week ahead of the sparrow-colored females.

During this week they have a rollicking good time. The air is filled with their tinkling, bell-like songs. Each one stakes out his little territory and defends it by frequent mad dashes at trespassers. As I look over this deliriously happy scene I'm sure there are a hundred Bobolinks but no, I count and there are only fifteen.

With the arrival of the females the courtship scene is even wilder, with males serenading females and chasing them all over the field, often three males pursuing one female or settling to the ground to woo a coy female hiding in the grass.

Soon they settle down to build their simple nest in a hollow on the ground. It is lined with fine grasses and surrounded by a little coarse grass. The five or six pale gray eggs splotched with brown are incubated twelve to fourteen days. The young leave the nest in eight to ten days but wander through the grass for several days before they are able to fly.

I have done as many as three hundred nest studies in one season, but never have I found a Bobolink nest. Ornithologists consider it one of the most difficult nests to find. But I believe in the old adage of "try, try again" so in early June I hunted a nest, walking through the waist-high bromegrass. Eventually I flushed a female who flew about 5 feet, then dropped down in the grass again. She is a tricky one, always running a considerable distance from her nest through the grass before flushing.

One observer reported that he located nests only after getting down on his hands and knees and crawling over a considerable area. Now Paul Harp is a patient man, generously granting me permission to wander over his land, but I had serious doubts that he would approve my wallowing down this prime grass in a crawling search.

So, standing up, I searched, backtracking through the grass, when suddenly I was surrounded by eleven male Bobolinks putting on a spectacular show, flying in wide circles around me, calling their bell-like notes not at all harsh and scolding because I suspect they knew I'd never find the well-hidden nest. It was my first observation of this particular behavior and a delightful performance it was!

Finally hot and tired, a vanquished foe, I retreated to my deck chair and spotting scope in the shade of the old deserted farmhouse to enjoy the magnificent panoramic view of the lake and bluffs in the distance. Once again all was peaceful in the Bobolinks' meadow world as each male flew back to his territory, perched on a swaying weed stalk, and sang "spink, spank, spink."

Carl Kurtz

Red-winged Blackbird

The most populous nesting bird in Iowa is the Red-winged Blackbird. Once the Dickcissel ran a close second and is still fairly common here although it is now blue-listed.

Numerous subspecies of the Redwing are widely spread over the North American continent except in arid deserts, high mountains, dense forests, and Arctic regions. Once they nested only in marshes, but with the draining of the marshes they have adapted to other habitats and now nest in shrubs or weeds in dry uplands or in hay fields.

I begin watching for them in March, as they are one of our earliest spring migrants. A joy it is to hear the glossy male perched on a post or shrub on territory singing his "oka-lee," awaiting the later arrival of the females.

On arrival she selects a nest site within the territory. She may be driven out or she may be accepted as a mate. His bows and postures are a ridiculous sight. He struts like a pea-cock during courtship, spreading his wings to display the brilliant red-and-yellow epaulets, spreading his tail and puffing out his body feathers. He also does a rough spiraling flight into the air and then drops down in a zigzag course.

In defense of his territory the male has a terrible scold and is very persistent. Years ago my mother reported that a male Redwing scolded her for three long hours as she hoed in the garden. I have also seen them rise above a hawk and strike at its back.

The substantial nest is built by the female with the attentive male looking on. Three days are spent building the outer basket and felting. During these three days it is not un-usual for the male to practice polygamy, al-though they are generally monogamous.

The five or six beautiful greenish blue eggs are heavily spotted and splotched with brown and incubated eleven to twelve days by the fe-male. Cowbirds often parasitize the nest, lay-ing one or two eggs in it. The nest is usually 2 to 3 feet up in a shrub or cattails. Althea Sherman earlier this century reported the

nests 15 to 22 feet up, but I have no records of high nests.

On hatching, the young are blind and help-less, with scarlet skin and a few tufts of buffy down. The growth in size each day is remark-able. On the second day feather sheaths of the primaries and secondaries show distinctly. By the ninth and tenth day they can fly and leave the nest. Sometimes the weakest one remains through the eleventh day.

By August flocks share common roosts in grassy sloughs or cattail marshes. The eco-nomic damage to crops is generally exag-gerated. Nevertheless, wholesale poisoning is carried out in the South, where roosts may contain a million Redwings and grackles. Relatively few overwinter in Iowa.

Carl Kurtz

Brown-headed Cowbird

The Brown-headed Cowbird is the only brood parasite in the eastern United States, habitu-ally laying her eggs in the nests of other spe-cies and never building her own nest.

Worldwide there are about eighty species of birds in several families that are completely parasitic at nesting time. Most are found in the Old World, but a few live in Central and South America. One other brood parasite, the Bronzed Cowbird, is found in southern Texas, New Mexico, and Arizona.

Ornithologists still dispute the causes of this evolutionary development, reaching no reasonable conclusion. We still have many unsolved puzzles in the scientific world.

Originally the cowbird was native to the grasslands of our midcontinent, usually associated with bison. As timber was cut and pastures established, millions of acres became prime habitat for this bird, with a resulting population explosion.

For far too long we have talked too glibly about the "balance of nature," not fully realizing how people have upset this balance. Now scientists are beginning to ask if the cowbird is also an agent of extermination, especially since their in-depth study of the cause of decline of the Kirtland's Warbler in Michigan. Between 1961 and 1971 the population of this little warbler found only in one region of Michigan dropped from 500 to 200 pair, a 60 percent drop, and the production of the young was very low. The study proved the cowbird to be the principal culprit.

A trapping program was initiated to control the cowbirds, and the response was dramatic. The warblers produced more than four young per pair of adults each year, laying to rest any doubts about the fertility of the eggs.

The female cowbird lays an average of twelve eggs per nesting season, and she does not discriminate. She tosses out an egg and then lays her egg in the nests of at least 214 species. Of these, 121 species have been known to rear cowbirds according to the cowbird authority, Herbert Friedmann. I added one last

species to the list when bluebirds reared a cowbird nesting in an oversize nest hole.

I believe the cowbird may be an important factor in the population decline of several of our small birds, including the Yellow Warbler, Bell's and Red-eyed vireos, and probably some of our smaller flycatchers here in Iowa. Fortunately, the cowbird ceases to lay eggs by July, so our state bird, the Goldfinch, escapes by nesting late in July or in early August. The Robin and the Catbird toss the cowbird eggs out of their nests.

B. Schorre/Vireo

Orchard Oriole

Everyone in Iowa is familiar with the brilliant orange-and-black Northern (Baltimore) Oriole, but few realize that we have another oriole. The Orchard Oriole, only 6 inches long, is slightly smaller than the Northern, not as brightly colored and far less numerous. In the summer of 1964 I banded seventy Northern Orioles but only one Orchard Oriole, all captured in mist nets set alongside a hedge of ripe mulberries a block west of my home.

A member of the blackbird family, the oriole adult male is black and chestnut, the female olive on the back and wings, with a greenish yellow throat and belly. The first-year male is one of our most beautiful birds, green all over but darker on wings and tail with a large black bib beneath his chin. A rare tropical bird is my first thought on seeing one.

There has been a steady population decline of the Orchard Oriole throughout this century. T. J. Barnes of Norwalk, born and raised near Lovilia, told me that every farm orchard had at least three to five nesting pairs in the early twentieth century.

Phil DuMont attributed part of the decline to grackle predation, but the increase of orchard insects and the corresponding increase in the use of arsenic sprays were probably greater factors. Beginning with World War II, the loss of habitat has been an important factor, with most of the old bluegrass pastures plowed under and all of the pioneer orchards long gone. Today orioles are endangered by the persistent pesticides and the herbicide spraying of highways and railroads. Some herbicides have contained up to 40 percent arsenic trioxide, sufficient to kill horses, and at least one contains highly toxic chlorinated dioxin contaminants known to produce birth defects in birds and mammals and believed to be the cause in Vietnam of a rare liver cancer, primary hepatoma, the result of defoliant spraying.

The Mockingbird field, an old bluegrass pasture with scattered shrubs and crab apple and red haw trees, is one of my favorite nest-snooping areas, with twenty-five to thirty species nesting there. Beginning with the March arrival of the Loggerhead Shrikes, "butcher birds" whose nests are complete in early April, the nesting season ends with the Goldfinches who never nest until near the first of August.

It is always mid-May before I see the first Orchard Oriole and early June before I find a nest, some years only one nest, some years two. To me the nest is even more interesting than that of the Northern Oriole. The Orchard Oriole weaves her lovely, globular hanging nest of long strands of green bluegrass.

It is not as deep as the Northern Oriole's nest, about 4 inches, and usually less than the outside diameter of 4½ inches, thin-walled with a contracted rim and a white, silky, wadded bottom. The great pioneer ornithologist-painter Alexander Wilson in 1832 found that one 13-inch strand of grass was hooked through and returned thirty-four times, winding round and round the nest.

It is truly a work of art. The green nest is so perfectly camouflaged among the green haw leaves that I am successful in finding a nest only when the bluegrass has cured to a beautiful yellow.

The first nest located in 1972 belonged to an adult black-and-chestnut male and his mate. It was 6 feet up in a slender haw tree and only 4 feet from a Northern Oriole nest 8 feet up in another slender haw. (Loss of elms is perhaps forcing the Northerns to accept lower nest sites.) This nest was depredated, and they moved across a ditch, a short distance east, to another slender haw only 30 feet from the gravel road.

On June 13 the nest contained two cowbird eggs and two oriole eggs. Herbert Friedmann, authority on host relations of the parasitic cowbirds, lists only eighteen records of the Orchard Oriole as host to that brood parasite, but I have rarely ever found a nest that did not contain cowbird eggs. On June 29, when I visited the nest, there was one baby bird with lemon-yellow skin, a little yellow fuzz, and big black eyes not open. The bluish white eggs marked with brown splotches and scrawls usually hatch in twelve days with the extreme being fifteen days. This one took sixteen days, and the other three eggs in the nest did not hatch at all—a possible indication of trouble, as both species have a high fertility rate. On July 3 there was a fresh cowbird egg in the nest, the first time I've ever known a cowbird to lay an egg in a nest containing a baby bird.

The third nest I located quite by accident as I was cutting across the field from the north Mocker nest to the east Mocker nest. I was being soundly scolded by a pair of kingbirds, when suddenly a very excited little green Orchard Oriole male started flying circles around a large haw tree and me. It took a few minutes to see the nest, too high to reach with my long-handled snooping mirror, and the tree was much too thorny for me to climb.

These orioles are friendly, sociable birds, getting along well with other species. Almost always their nest is near a nest of the pugnacious kingbirds, undoubtedly receiving some protection thereby. It is usually within 100 feet of a Mocker nest, and the Mocker also is a good fighter.

The Orchard's voice is unlike that of the Northern Oriole, not as loud and rich but very pleasant and somewhat Robin-like. Occasionally, it sounds like "look here, what cheer, what cheer, whip up, what cheer, we yo," but more often it defies description, so I just listen and enjoy it.

One evening I sat in the car near a nest hoping to tape the adult male's song. He sang not at all, but he gave a scolding note "cheep, cheep" again and again for over half an hour until, defeated, I drove away.

The Orchard Orioles spend fully nine months of the year in the tropics of Central America, chiefly in Guatemala and Honduras. George K. Cherrie (who grew up near Knoxville), an early twentieth-century ornithologist-collector for the Field Museum, Chicago, and the American Museum, New York, reported them returning to Costa Rica by July 31. I seldom see them after mid-July. Woodward H. Brown, in "An Annotated Checklist of the Birds of Iowa," lists the earliest date as May 1 and the latest date as August 29.

Carl Kurtz

Northern Oriole

Changing the name of our beautiful Baltimore Oriole to Northern Oriole infuriates me, and I vow that I will never, never call that bright and glowing bird by such an icy-cold name. I shall continue to call that gay charmer of my childhood Baltimore Oriole to the day I die!

It was always early May when Mother announced "the Baltimore Oriole has returned" and Sis and I rushed out to see that gaudy bit of feathers, to hear it sing, and to begin the watch for the nest.

Every farmyard had a nesting pair, seldom more as they are fierce territorial defenders. Here in town our ancient elm had a nest each of its last twenty-five years and doubtless had one for the preceding seventy-five years of its hundred-year life span.

I missed their nesting in Georgia, as they do not nest below the Fall Line, but enjoyed their spring migration in mid-April and fall migration in August with one occasionally overwintering. The majority winter through Central America into Colombia and Venezuela.

In 1967 my neighbors Phil and Dorothy Proffitt had a dying elm removed from their backyard. We were all distressed because a pair of orioles was busily feeding young in a hanging nest in the doomed tree. After considerable discussion, we asked the man in the high bucket to cut off the nest branch, swing the bucket across the driveway to the Hubert Isenbergs' huge silver maple, and wire the nest branch securely to a maple branch. The parent birds watched and scolded all the while but responded immediately to their babies' cries and resumed feeding with no hesitation at all, whereupon the dozen human onlookers heaved sighs of relief.

In the summer of 1964 I banded seventy orioles captured in mist nets set alongside young mulberry trees in my neighbor's acreage. With the loss of our elms, the orioles are hard-pressed for nest sites and the population has declined here.

Now I find them nesting much lower, 9 feet up in a haw tree, 10 to 12 feet up in saplings, and even one in a weeping willow tree. Most of the tree-lined ditches on farms have been bulldozed, and the tree-bordered country roads are no more.

For obvious reasons (I am not a climber) I have banded very few oriole nestlings, but there's one nestful I'll never forget. Our local scoutmaster Lester Hancock suggested one summer that we band the nestlings at least 15 feet up in a City Park tree. And we did!

An extension ladder was held straight up in the air while his son John climbed to the nest, then lowered the nestlings in a can attached to a rope. I quickly banded them, then John pulled up the can and restored the babies to their nest. The entire operation took less than ten minutes.

The Baltimore Oriole is the most skillful artisan of any North American bird in constructing its woven, hanging-pouch nest of string and bits of cloth. Horsehair, long a favorite, is now seldom available. An acceptable substitute is 6- to 8-inch lengths of binder twine dipped in hot water and then separated into fine strands.

The four to six grayish white eggs marked with black, brown, or lavender scrawls, usually on the larger end, are incubated twelve to fourteen days. The nestlings usually remain in the nest about two weeks and are fed by both parents on insects, mainly caterpillars. The adults also are insectivorous but are very fond of fruit, Juneberries (shad), mulberries, blackberries, and grapes. I always leave part of my Concord grapes for them. Nectar of flowers is a favorite. At the feeding station, suet syrup and half an orange are relished.

Carl Kurtz

Pine Grosbeak

To bird-watchers, the excitement of winter depends on the appearance of the northern finches and other boreal species. In 1975, 80 Snowy Owls were sighted in thirty-one Iowa counties—and thousands of Purple Finches. In 1973–74 it was the western Montana species, Clark's Nutcracker, with 45 seen in twenty-nine Iowa counties. There was a fair flight of Red Crossbills in 1972–73 and a major flight of Redpolls in 1971–72. Fred Kent reported 150 Redpolls seen on the 1975 Iowa City Christmas Bird Count.

One observer of Evening Grosbeaks in 1976, Marvin Pottorf in Sheldon, informed me via long-distance telephone that they not only had Evening Grosbeaks but also 8 to 12 Pine Grosbeaks in the cemetery pines across the street from their home during Christmas week. They studied them with binoculars, a positive identification.

That was exciting news as these rosy red, robust, Robin-sized winter finches are the rarest of the winter finches, seldom reaching Iowa and then only the northern half of the state. (Roy Oliver observed them once in Mount Pleasant and Helen Peasley once in Des Moines.) Because they so seldom reach

Iowa, this species was removed from the Iowa Ornithologists' Union field checklist in 1975.

I immediately alerted Lorraine Wallace of rural Spirit Lake so she could make inquiries in her column for two county newspapers. She is the IOU bird reporter for northwest Iowa. Her reports and stories are published in *Iowa Bird Life*.

This rosy finch with the heavy black grosbeak (shaped like the Cardinal's beak), dark wings with two white wing bars, and a dark tail is easily identified. Remember, it is Robin-sized. Do not confuse it with the sparrow-sized Purple Finch, White-winged Crossbill, or Red Crossbill. The difference in size is obvious. The Pine Grosbeak females and young birds are gray with olive to rusty gold heads, dark wings, white wing bars, and dark tails.

Pine Grosbeaks do not make regular fall and spring migrations. However, there are invasion years. A few years ago Dr. J. Murray Spiers published data showing peak populations occurring at intervals of five to six years in the Toronto, Ontario, area. Ludlow Griscom reported on ten marked flights in ninety-six years in the New York City area, the last one in the winter of 1903–04, then none for eighteen years. Massachusetts had at least four great invasions during the winters of 1869–70, 1874–75, 1892–93, and 1903–04.

In recent years irruptions occurred in 1961 and 1965, with flights heaviest in Canada and the Pacific states in 1965, while in the states from Michigan and Ohio east the flight was larger in 1961, with New York reporting 1,842 birds. Minor flights occurred in 1969–70 and 1971–72, with a major irruption in 1972–73. Very few ever reached as far south as Iowa, although Minnesota recorded several hundred. It was not severe winters that drove these birds south, some were mild and open winters. Major invasions appear to be governed entirely by the food supply.

A half-century ago Iowa's own Ira N. Gabrielson, onetime chief of the U.S. Fish and Wildlife Service, analyzed their winter food and found it to be 99.1 percent vegetable matter, with the summer food 16.7 percent animal matter, grasshoppers, ants, spiders, and caterpillars.

Besides pine seeds, a wide variety of seeds and fruits are eaten. Favorites are crab apples, mountain ash berries, maple buds, and pine seeds, but many others are eaten—cranberries, barberry, hawthorn cedar, rose, blackberries, snowberries, beechnuts, acorns, coniferous seeds and buds, and mountain holly. They are very fond of sunflower seeds, hemp, burdock, ragweed, and lamb's-quarter seeds.

In 1955 the late William Youngsworth of Sioux City observed the birds splitting the seed pods and extracting the seeds of Persian lilacs. During the 1954–55 invasion in West Virginia, Maurice Brooks found that the birds fed heavily on frozen apples and other fruit, also maple, white ash, tulip tree, wild grape, dogwood, green berries, sumac, privet, and bush honeysuckle berries.

They are quite tame and unafraid of people, allowing close approach. When feeding, they often keep up a whistled conversation. The warning note is a whistled "carey." Occasionally, they sing in winter, a sweet melodious carol, loud and distinct, full of trills and warbles.

This lovely bird is circumpolar, making its summer home in the coniferous forests of Alaska, northern Canada, northern Europe, Asia, and Siberia, rarely in the northern parts of New England and some of the more western states. It nests as far north as the tree limit in northern Canada, with at least a few nest records in northern Maine and northern Wisconsin. The nest is compactly built, with a foundation of small twigs and roots and lined with fine rootlets, grasses, and lichens at a height of 2 to 15 feet. The beautiful eggs are blue-green, speckled or splotched with brown, black, or purple markings, and are incubated about thirteen days. Both parents feed the nestlings during the twenty-day nest life and continue feeding them for several days after fledging.

Larry Stone

Purple Finch

What a joy it was to have a flock of Purple Finches attracted to my feeders in late March. Immediately I ordered more sunflower seeds, knowing full well that those on hand would fast disappear.

Sparrow-size, the male is rosy red, deepest on the head and rump, "like a sparrow dipped in raspberry juice." This rosy red color is supposedly the biblical purple, hence the name. The female is heavily striped brown with a white line over the eye.

Their sweet warbling song is a joy. The entire flock formed a chorus warbling from the top of my old apple trees. Every song seemed different. Bird-song analysts say this spring song consists of six to twenty-three notes, very rapid, averaging seven notes per second. They also have two other distinct songs, the "territory" or nest song after pairing and a very unusual "vireo" song heard rarely in early spring and less often in late fall.

Migration is erratic, mostly north-south but sometimes east-west. According to Banding Office records, 4,700 were banded per year prior to 1939, which was an invasion year with 21,592 banded.

Sometimes one locality has an invasion with few seen elsewhere in the state. From January 1 to May 1, 1964, Charles and Darleen Ayres banded 1,465 in Ottumwa, but only 42 were banded elsewhere in Iowa. Most were trapped on the Ayres's front porch and the rest in their backyard. About 186 of the banded finches were retrapped that winter.

Twenty-eight birds were retrapped in Ottumwa in later years, the oldest was one banded February 15, 1964, which returned March 22, 1969. (Longevity average is three to four years with extremes of six to ten years.) Eight of the banded birds were caught or found outside Iowa: 4 in St. Paul, Minnesota, 1 in Little Falls, Minnesota, 1 in Marrilton, Arkansas, 1 in Midland, Michigan, and 1 in St. Vitale, Manitoba, Canada.

In winter they are gregarious and friendly, traveling in flocks. Considerable squabbling goes on at the feeding trays, but they have their own ERA and the peck order is not dominated entirely by males. The many bossy females viciously drive males off the feeders. I solved the problem by maintaining five sunflower feeders so all had a chance to eat. In Georgia, our Purple Finches arrived around Christmas shortly after the Cedar Waxwings. They were scavengers in our yard, sweeping up the ligustrum seeds in the waxwing droppings, neatly crackling the seeds and eating the kernels. In the Georgia woods, I watched them eat seeds of the sweet gum burs and the canelike seed pods of the tulip tree. They like the seeds in the sycamore balls here, and often the blowing fuzz reveals their presence. Weed seeds of all kinds are relished. In spring they eat the buds and blossoms of apple trees, feeding on the stamen and pistils. (Cedar Waxwings also enjoy apple blossoms.) During the summer they eat many insects.

They nest across the northern states and in Canada but not in Iowa. Two years of my childhood were spent on a Wisconsin farm where Purple Finches nested in the evergreen forest that we reached by a walk through a pasture across a brook with tiny trout, pausing to gather waxy yellow buttercups, blue columbine, and tiny white sweet-scented violets.

The neat nest of grass and roots lined with horsehair contained four to five blue-green eggs spotted with brown. Later when the young were out of the nest and making their "pee, wee" clamor to be fed, we spent many mornings picking delicious wild red raspberries.

Carl Kurtz

House Finch

The so-called Hollywood Finch is actually the small House Finch native to the Pacific slopes of central and southern California north to British Columbia and east to Colorado, Wyoming, and West Texas. It had never extended its range east of Colorado except for a few sighted in western Nebraska. The 1981 Christmas Bird Count listed four at Grand Island, sixty-four at North Platte, and forty-one at Scottsbluff.

In 1940, some enterprising cage-bird dealers on Long Island, New York, received a shipment of House Finches. As U.S. Fish and Wildlife Service agents were about to make arrests, the dealers released the finches on western Long Island.

Thus began an opportunity for ornithologists to document the range extension of a small songbird from a known point of origin. Through the National Audubon Society's Christmas Bird Counts, the winter range was reported annually. The count for 1947– 50 reported 35, 104, 36, and 37 House Finches, and it was estimated that 225 to 250 birds comprised the total breeding population on western Long Island in 1951. From that point of origin, Mundinger and Cale estimated in

American Birds (July 1982) that the 1978 winter range was 169,120 square miles, almost a thousandfold increase in just thirty years, and that this expansion is accelerating.

At first, the finches expanded most rapidly along the coast, up and down the major river valleys, and toward the southwest. In 1967–1971 they moved up the valleys of the Connecticut and Hudson rivers. In 1976–1979 they expanded their winter range very rapidly down the Ohio River, with one bander banding more than two hundred in Loraine, Ohio.

The House Finch is firmly established as a breeding bird in all the northeastern states and is spreading rapidly into the Midwest with a few sightings in Illinois, Indiana, Minnesota, and Ontario. Winter records into the Carolinas and as far south as central Georgia have been made.

So I had been eagerly anticipating a first report in Iowa. On July 20, 1982, Wallace Jardine of Pocahontas called to tell me he had a male House Finch at his feeders for one day but did not get a photograph. At least two experienced birders must identify a rare species to document it or else a photograph must be taken, so this sighting could not be listed as a documented record.

The next day, an amateur birder with one year's experience, Marcia Steck of Perry, phoned to tell me she was quite sure she had had a male House Finch at her feeders for the past month. The nearest experienced birders were Gene and Marilyn Burns of Jamaica, and they identified the bird as a male House Finch but failed to net the bird and band it. I also phoned the president of the Iowa Ornithologists' Union, Dr. Ross Silcock of Malvern, and on July 26 he, Dr. Tom Kent of Iowa City, and Jardine visited the Steck residence, photographed the bird, and made a positive identification—the first official record of a House Finch in Iowa. Another male was netted, photographed, and banded by Don Johnson in Ottumwa on July 26.

The House Finch is a sparrow-sized bird with a short, stubby bill. The male has a bright poppy-red head, throat, and rump, with streaked sides. The similar Purple Finch is rosy red on the head, throat, breast, sides, and back, with no streaking on the sides or belly. The female House Finch has finer streaking on the sides, belly, and head. The female Purple Finch has heavier streaking on the breast and belly and a distinctive white line through the eye.

Be sure to check your flock of Purple Finches, although the House Finches prefer to migrate farther south in the winter. However, a lone Purple Finch in summer is suspect and should be checked carefully.

House Finches are great singers, and Marcia remarked that her bird sang constantly.

Preferring urban areas, they are very gregarious. In some areas of California, flocks are almost as numerous as House Sparrows.

Chiefly seedeaters, House Finches will scavenge bread crumbs and table scraps. Sometimes they feed on apricots, plums, and cherries in orchards. Apparently they do not eat insects but are very fond of suet.

The nest sites are varied, including tree branches, eaves, tin cans, and bird houses. The three or four bluish white eggs speckled with brown are incubated twelve to fourteen days, and the nestlings remain in the nest fourteen to sixteen days.

House Finches invaded Iowa in the early 1980s. The first recorded nesting was in 1986. By 1989 they were being reported across most of the state.

Larry Stone

Red Crossbill

Strange indeed are the various types of beaks of the nearly 9,000 birds of the world. The primary purpose of a bird beak is for eating, but the beaks come in all shapes and sizes. There are long bills, short bills, flat bills, down-curved and up-curved bills, hooked bills, spearlike bills, and tube noses of the petrels through which salt is excreted.

A very odd bill is that of the Roseate Spoonbill. The broadened bill end is full of nerve endings to feel for animals when grubbing in the mud. This is also true of the long slender bill of the woodcock, an aid when probing for insects and worms in the soil of moist woods. Another odd beak is that of the Black Skimmer, whose lower mandible is longer than the upper. Feeding occurs in flight when the lower mandible cuts the water and strikes food below the surface.

The seedeaters have short strong bills—strong enough to crack seeds and also to raise blood blisters on the hand of the careless bander.

But to me the strangest bills of all are those of the crossbills. The ends of their two mandibles cross in opposite directions, allowing them to extract seeds from the cones of ever-green trees. The tip of the upper mandible is always greatly prolonged beyond the tip of the lower. And the birds are either "right-handed" or "left-handed" in opening cones, according to which way the mandibles are crossed.

On March 9, 1984, I received a long-distance call from J. B. Garbison of Lucas asking if it could have been a Scarlet Tanager at his Niger seed feeder that morning. I re-plied that it was highly unlikely since the Scarlet Tanager is an insect eater and does not return from the tropics until mid-May.

Once before I had a winter report of a Scarlet Tanager, and I knew immediately that Mr. Garbison's bird was a male Red Cross-bill. The color is usually more brick-red, but sometimes the color of adult males is bright scarlet with dark wings and tails.

Mr. Garbison's bird did not master the tech-nique of removing Niger seeds, but several years ago a White-winged Crossbill learned to tweezer Niger seeds from the tiny holes of a Goldfinch feeder in Pella and spent the win-ter there. This brought great joy to the many birders who came to see this rare finch.

Mr. Garbison's home is only half a mile from fifty-year-old pines in Stephens Forest, and we speculated that the bird had probably wintered there.

This was the first Red Crossbill report I had received in 1984. Never too plentiful, they are found mostly in the northern half of Iowa. However, Mrs. John Neighbors of Chariton has in past years observed them in mature evergreens in the Chariton cemetery. And the earliest ever fall record was August 10, 1956, when a small flock was seen in ever-greens on the Simpson College campus in Indianola.

Found mostly in coniferous forests of the far north during the breeding season, a few breed in the Great Smoky Mountains, in west-ern mountains, and in the mountains of Mex-ico. They have no fixed season for breed-ing—sometimes in January or February, most commonly in late winter and early spring, less often in September and October, and still more rarely in May, June, and July. It is prob-ably triggered by an abundant supply of cones, as the babies are fed by regurgitation.

Crossbills eat the seeds from the cones of various pines, firs, spruces, hemlocks, and larches, but they also eat seeds of birches, al-ders, and ragweed and other weeds and the buds of birches and other deciduous trees. In season they also eat insects and caterpil-lars. Their appetite for salt is well known, and many are killed on salted highways in Canada.

Both species are circumpolar and are no-torious nomads. One year there may be hun-dreds, and then none may appear for several years.

Carl Kurtz

Common Redpoll

Not often do we have the pleasure of the company of those erratic winter visitors, the Common Redpolls. Bearing a very appropriate scientific name, *Acanthis flammea*, which I choose to translate roughly as "flower of flame," they would seem to belong to a tropical clime of brilliant colors.

Instead, they are birds of the far north. The edge of the tundra marks their northern limits and there they spend most of the year, rarely wandering in winter as far south as the fortieth parallel.

They are little brown-streaked birds, 5 inches long; the distinctive marks of the male are the bright rosy red cap, black chin, pink breast, and pinkish rump. The female lacks the pink breast. The tail and wings are brownish with two white wing bars. The yellow bill is short and conical, typical of the seed-eating finches.

In mid-January of 1972, Bill Criswell (Des Moines Audubon Society) observed a flock of one hundred or so in weed fields along the north shore of Lake Red Rock just west of Robert's Creek Lake.

A week later Bill and I found a flock of thirty-five near the long sandbar along the south section of old Iowa Highway 14. Later we found them eating weed seeds on the ground along the north end of old Highway 14. They were quite "tame" and so very beautiful, allowing us to approach to within 15 feet. This "tameness" is characteristic of many far-northern birds—no fear of humans.

They are restless little birds, feeding in one spot for several minutes, then all take wing and move to another feeding site. I have never heard an alarm note sounded, it's just as if they have a common mind—all take off at once.

Their flight is undulating, very like the Goldfinch's, and unless I hear "cheet cheet" or see the pink color, I cannot be sure of their identity on the wing.

Redpolls are circumpolar, found in northern Scandinavia, northern Russia, Siberia, Alaska, northern Canada, Newfoundland, and Labrador. Migrating a little south, they winter in all those countries, also in France, Italy, Yugoslavia, Turkey, China, Korea, north Japan, and in our northern states.

Redpolls, considered Arctic or subarctic permanent residents, overwinter the farthest north of any small North American bird. At temperatures as low as −60 degrees C, how can such a small bird weighing only 12 to 14 grams survive?

That question was answered by Dr. W. S. Brooks (now of Ripon College). In research on redpolls at the Alaska Arctic Research Laboratory and the University of Illinois he concluded: They select high-calorie food, such as birch seeds, over others. In addition to their crop, they have a special storage pouch, a croplike esophageal diverticulum which is filled with extra food just before nightfall for needed energy overnight. They feed later in the evening, often in darkness, and they fly to the food trees before daylight. They have many down feathers during winter, giving them greater insulation. They can increase digestive efficiency at low temperatures.

I have banded only three redpolls, all males, on March 12, 1963. I had observed four bathing in the icy waters of a little stream a few days earlier. Disregarding the cold, they were having a glorious time splashing water, a really vigorous bath. (Other observers have reported snow bathing.)

I set a mist net across the ditch but failed to capture them. After observing them feeding on lamb's-quarter (*Chenopodium album*) seeds—I stood 5 feet away, they paid no attention to me—I set my net alongside the weeds and captured three of the four.

The exquisite colors and soft plumage can only be fully appreciated in the hand. Normally very peaceable, they displayed a little temper, pecking at each other in the holding cage, and one pecked my hand.

The redpolls nest even on the tundra if they can find driftwood for a nest site, but the main habitat is the semibarrens, where they nest in dwarfed spruces or in willow and alder thickets. The female selects a male, acts aggressive toward him, and soon the pair bond is formed. The males then sing constantly, but there are no territorial battles as the nests are often quite close together.

The nest is in the crotch of an alder or willow or across a spruce branch and has a base of sticks and a loose cup of twigs and grass. The thick, deep inner cup of ptarmigan feathers provides warmth for eggs and nestlings and almost hides the female.

The four to five eggs are blue-green, thickly sprinkled with purplish spots on the larger end. The females are close sitters,

and incubation time is eleven days. Roland Clement (in Bent's *Life Histories*) in studies in northern Quebec reports the young are hatched naked except for a bit of down on tail and wings.

The skin is so translucent that food can be seen in the gullet, and the blood vessels give the skin an orange tinge. They grow rapidly, eyes open by the fifth day, and they are able to fly enough to leave the nest by the twelfth day. The long days of the subarctic summer provide twenty hours of daylight. Grinnell at Churchill, Manitoba, reported that adult birds were observed carrying insects to nestlings from 3 A.M. to 10:30 P.M., and Walkinshaw farther north in Alaska reported around-the-clock activity.

The only other redpoll found on this continent is the Hoary Redpoll. In appearance it differs from the Common Redpoll in having an unstreaked white rump and a paler back, a frostier look.

It too is circumpolar. It differs very little in habits except that it breeds farther north in the tundra and it also winters farther north, rarely reaching the northern states and never reaching Iowa.

The southern edge of the breeding range of the Hoary overlaps the northern edge of the breeding range of the Common, and they do interbreed. Dr. Brooks is of the opinion that there is but one species of redpoll. The two are subspecies, with the Hoary being adapted to lower temperatures, more Arctic.

American Goldfinch

In selecting a state bird our pioneer ancestors could not have made a better choice than the American Goldfinch. Often called the Wild Canary by old-timers here—in the Georgia Okefenokee Swamp it is called the Kee-Hee Bird—its disposition is every bit as sunny as its coat.

Its song is sweet and high-pitched. Its cheerful call notes "kee, hee" or "perchicka-ree" are often heard during its undulating flight. Even the begging notes of the hungry nestlings have a musical tone. I have never heard a scolding note of any kind.

Thanks to neighbor George Dyer's "bee garden," it has been my pleasure to have daily visits by Goldfinches all through July, August, and September. (The "bee garden" is really my old vegetable garden which George planted to anise hyssop last spring.)

The 3-foot-tall plants are covered with spikes of lavender blossoms high in nectar and very attractive to George's eight hives of bees. Many butterflies and moth species are also attracted, so it is truly a garden of color plus sound and motion.

Anise hyssop is a member of the square-stalked mint family, although the leaves have an anise odor. It is not the nectar that attracts the Goldfinches, rather it is the milky seeds. The adults feast on them, and their nestlings are fed not on insects but on regurgitated milky seeds.

This little finch is a seed eater; probably 95 percent of its diet is vegetable and only 5 percent insects. All kinds of seeds are relished, but the favorites are dandelion, thistle, evening primrose, and sunflower. In 1963 I banded seventy-five Goldfinches captured in mist nets set alongside sunflowers in my neighbor's garden. I have also observed them eating the seeds in an Osage orange crushed by a passing car, and in Georgia they ate the seeds of sweet gum and tulip trees.

The Goldfinch is the only Iowa bird that flits the summer away, waiting till late July or early August to begin its nesting activities—then it is assured a plentiful supply of milky seeds to feed the young.

The courtship is a merry affair. Often both the male and female fly in great circles, with the male singing all the while. Again it may be only the male in courtship flight, with slow wingbeats flying on an even keel unlike his usual undulating flight.

The beautiful nest is strong and thick-walled, constructed of plant fibers and nearly always lined with thistledown. I once found a nest built largely of ripe red corn silks but lined with thistledown, a very pretty nest. The female spends about six days building her nest, usually in the fork of a small tree or big thistle 4 to 9 feet up. Most of the nests I've studied were in box elders or bull thistles along the CB & Q and Rock Island railroads.

The four to six pale bluish white eggs are incubated twelve to fourteen days by the female being fed by the male. She is a constant sitter, leaving the nest infrequently. The nestlings remain in the nest thirteen to seventeen days and are fed by both parents. There was a light frost in the lowlands in mid-September one year while I was watching a nest of three quite naked little Goldfinches, located along the Rock Island tracks near Coal Creek. I rushed out the next afternoon sure that the baby birds had perished from the cold, but I need not have worried. They were chipper as could be, well protected by the thick walls of the nest as well as the heat of the mother bird's body. I banded the three on September 20.

Most nestlings are banded in August, but even our August nights are quite cool, so the warm nest is absolutely essential to the survival of the species.

Everyone knows the yellow-and-black male Goldfinch, but few recognize the dull-colored female with the olive-yellow body, brown tail, and wings marked with two conspicuous white wing bars. Fewer still realize that the male acquires the same dull plumage during the winter months.

While the majority migrate south as far as Mexico in winter, a fairly large population remains in Iowa, with Christmas counts ranging from fifty in Red Rock Refuge to a high of two hundred in Davenport. .

The northward migration in spring is generally a leisurely affair with some arriving by mid-March, but there is no hurry, no pressure to nest early. In Georgia I once saw a flock on May 8 that numbered in the thousands, all balancing gracefully on ripening oats, picking at the milky grains. It was such a beautiful sight that I simply forgot my errands, just sat and looked and listened.

Larry Stone

Evening Grosbeak

If one cold wintry day you look at your bird feeder and see an oversize "Goldfinch," don't call your veterinarian, the bird does not have a pituitary imbalance, and don't call your oculist, you do not have an eye problem. You are looking at an Evening Grosbeak, a nomad from the north, cousin of the Cardinal and the Rose-breasted Grosbeak.

A chunky bird, Starling-sized, with a heavy white beak, the male has a yellow body, some black on the head, black wings with big white wing patches, and a black tail. The female is gray with yellow on the nape, sides, and upper tail coverts, a black tail, and wings spotted with white.

They are very handsome birds and a joy at the feeders, albeit a costly joy. One friend in the Southeast fed a ton of sunflower seeds during one winter and considered it money well spent. Some years are irruption years with hundreds and the next year none.

This is another species that has extended

its winter range; originally seen only in the West, by 1890 it moved eastward to southeastern Massachusetts, then began drifting southward as far as North Carolina and Rome, Georgia. During the winter of 1968–69 hundreds were banded at Warner Robins, Georgia, on the coastal plains 25 miles south of the Fall Line.

Evening Grosbeak authorities say these birds feed on the winged seeds, or samara, of the box elder trees (the softest of our maples, called Manitoba maples in Canada) all fall, and when these and weed and fruit seeds are exhausted they finally resort to sunflower-seed feeders.

Every fall and winter I go peer into box elders, but never have I seen an Evening Grosbeak there. However, I once watched them feeding in the top of a tall sugar berry tree on the campus of Mercer University, Macon, Georgia. Newton wildlife photographer Herb Dorow took pictures of a flock of fifteen feeding on weed seeds in one of the Red Rock Refuge fields during November.

They have great adaptability, changing their diet with the changing seasons. "Budding," the eating of tree buds of maples and various other trees, and the drinking of sugar-maple sap dripping from the ends of twigs are the spring feeding pattern. In maple-sugar country they are called Sugar birds, perhaps a more fitting name as they definitely are not active in the evening. Rather, they roost quite early in the afternoon. The misnomer resulted when their discoverer disturbed the flock as he set up evening camp; naturally the birds voiced their resentment.

Many winters they remain in the north if the food supply is adequate. According to Canada Evening Grosbeak authorities J. Murray and Doris Spiers, they are well equipped to withstand bitter cold, being insulated with eiderdown, a thick gray down under the contour feathers, and they have their own "snow tires," ridged cushions on the underside of the feet and toes to enable them to grip icy branches.

Highly gregarious, the flock keeps in contact during the slightly undulating flight with a "pete" or "p-teer" note. Although the flocking note is best known, they have a number of other calls. The flock sometimes joins in a chorus song, a little like the Purple Finch's "chip-choo-wee." In March the males sing a similar whisper song, then as the breeding season approaches, both sexes sing the same song full voice.

It never has been my good fortune to be in the North Woods at the right time to observe the nesting of these beautiful birds. Others report the nest of sticks lined with fine roots and lichens (but no feathers) built in deciduous and coniferous trees usually 25 to 55 feet up.

The three to five blue-green eggs spotted with brown are incubated twelve to fourteen days by the female. Both parents feed the young a diet of insects, with spruce bud-worms a favorite. The nestlings' begging food-call is "see-see-see," and the fledglings' food-call is "bee? bee, bee?"

Their distribution in summer is fairly widespread, with a decided preference for cooler regions. The eastern race goes into Canada to nest, the western race into the mountains of Montana and British Columbia, and the Mexican race into the Chiricahua Mountains in southeastern Arizona and the mountains of southeastern Mexico.

Checklist of Iowa Birds

Species names and status information are from the "Official Checklist of Iowa Birds, 1991 Edition," by Thomas H. Kent and Carl J. Bendorf, *Iowa Bird Life* 61, no. 4 (Fall 1991), published by the Iowa Ornithologists' Union. Species names follow *The AOU Check-list of North American Birds*, 6th edition, 1983, and supplements through 1991. Family names and taxonomic order are from the *ABA Checklist: Birds of the Continental United States and Canada*, 4th edition, 1990.

Species	*Status*
LOONS	
Red-throated Loon	casual
Pacific Loon	casual
Common Loon	regular
GREBES	
Pied-billed Grebe	regular
Horned Grebe	regular
Red-necked Grebe	regular
Eared Grebe	regular
Western Grebe	regular
Clark's Grebe	accidental
PELICANS	
American White Pelican	regular
Brown Pelican	accidental

Species	Status
CORMORANTS	
Double-crested Cormorant	regular
DARTERS	
Anhinga	accidental
FRIGATEBIRDS	
Magnificent Frigatebird	accidental
BITTERNS AND HERONS	
American Bittern	regular
Least Bittern	regular
Great Blue Heron	regular
Great Egret	regular
Snowy Egret	regular
Little Blue Heron	regular
Tricolored Heron	accidental
Cattle Egret	regular
Green-backed Heron	regular
Black-crowned Night-Heron	regular
Yellow-crowned Night-Heron	regular
IBISES AND SPOONBILLS	
Ibis species	regular
White-faced Ibis	casual
Roseate Spoonbill	accidental
STORKS	
Wood Stork	accidental
WHISTLING-DUCKS, SWANS, GEESE, AND DUCKS	
Tundra Swan	regular
Trumpeter Swan	extirpated
Mute Swan	regular
Bean Goose	accidental
Greater White-fronted Goose	regular
Snow Goose	regular

Species	Status
Ross's Goose	regular
Brant	accidental
Canada Goose	regular
Wood Duck	regular
Green-winged Teal	regular
American Black Duck	regular
Mallard	regular
Northern Pintail	regular
Blue-winged Teal	regular
Cinnamon Teal	regular
Northern Shoveler	regular
Gadwall	regular
Eurasian Wigeon	accidental
American Wigeon	regular
Canvasback	regular
Redhead	regular
Ring-necked Duck	regular
Greater Scaup	regular
Lesser Scaup	regular
Common Eider	accidental
King Eider	accidental
Harlequin Duck	accidental
Oldsquaw	regular
Black Scoter	regular
Surf Scoter	regular
White-winged Scoter	regular
Common Goldeneye	regular
Barrow's Goldeneye	accidental
Bufflehead	regular
Hooded Merganser	regular
Common Merganser	regular
Red-breasted Merganser	regular
Ruddy Duck	regular
AMERICAN VULTURES	
Black Vulture	accidental
Turkey Vulture	regular

Species	Status	Species	Status
KITES, HAWKS, EAGLES, AND ALLIES		Sora	regular
Osprey	regular	Purple Gallinule	accidental
American Swallow-tailed Kite	extirpated	Common Moorhen	regular
Black-shouldered Kite	accidental	American Coot	regular
Mississippi Kite	accidental		
Bald Eagle	regular	**CRANES**	
Northern Harrier	regular	Sandhill Crane	regular
Sharp-shinned Hawk	regular	Whooping Crane	accidental
Cooper's Hawk	regular		
Northern Goshawk	regular	**PLOVERS AND LAPWINGS**	
Red-shouldered Hawk	regular	Black-bellied Plover	regular
Broad-winged Hawk	regular	Lesser Golden-Plover	regular
Swainson's Hawk	regular	Snowy Plover	accidental
Red-tailed Hawk	regular	Semipalmated Plover	regular
Ferruginous Hawk	accidental	Piping Plover	regular
Rough-legged Hawk	regular	Killdeer	regular
Golden Eagle	regular		
		STILTS AND AVOCETS	
CARACARAS AND FALCONS		Black-necked Stilt	accidental
American Kestrel	regular	American Avocet	regular
Merlin	regular		
Peregrine Falcon	regular	**SANDPIPERS, PHALAROPES, AND ALLIES**	
Prairie Falcon	regular	Greater Yellowlegs	regular
		Lesser Yellowlegs	regular
PARTRIDGES, GROUSE, TURKEYS, AND QUAIL		Solitary Sandpiper	regular
Gray Partridge	regular	Willet	regular
Ring-necked Pheasant	regular	Spotted Sandpiper	regular
Ruffed Grouse	regular	Upland Sandpiper	regular
Greater Prairie-Chicken	accidental	Eskimo Curlew	extirpated
Sharp-tailed Grouse	extirpated	Whimbrel	casual
Wild Turkey	regular	Long-billed Curlew	accidental
Northern Bobwhite	regular	Hudsonian Godwit	regular
		Marbled Godwit	regular
RAILS, GALLINULES, AND COOTS		Ruddy Turnstone	regular
Yellow Rail	regular	Red Knot	casual
Black Rail	accidental	Sanderling	regular
King Rail	regular	Semipalmated Sandpiper	regular
Virginia Rail	regular	Western Sandpiper	regular

Species	Status	Species	Status
Least Sandpiper	regular	Ivory Gull	accidental
White-rumped Sandpiper	regular	Caspian Tern	regular
Baird's Sandpiper	regular	Common Tern	regular
Pectoral Sandpiper	regular	Forster's Tern	regular
Sharp-tailed Sandpiper	accidental	Least Tern	regular
Dunlin	regular	Black Tern	regular
Curlew Sandpiper	accidental		
Stilt Sandpiper	regular	AUKS, MURRES, AND PUFFINS	
Buff-breasted Sandpiper	regular	Thick-billed Murre	accidental
Ruff	accidental	Ancient Murrelet	accidental
Short-billed Dowitcher	regular		
Long-billed Dowitcher	regular	PIGEONS AND DOVES	
Common Snipe	regular	Rock Dove	regular
American Woodcock	regular	Mourning Dove	regular
Wilson's Phalarope	regular	Passenger Pigeon	extinct
Red-necked Phalarope	regular		
Red Phalarope	accidental	LORIES, PARAKEETS, MACAWS, AND PARROTS	
		Carolina Parakeet	extinct
SKUAS, GULLS, TERNS, AND SKIMMERS			
Pomarine Jaeger	accidental	CUCKOOS, ROADRUNNERS, AND ANIS	
Parasitic Jaeger	accidental	Black-billed Cuckoo	regular
Long-tailed Jaeger	accidental	Yellow-billed Cuckoo	regular
Laughing Gull	accidental	Groove-billed Ani	accidental
Franklin's Gull	regular		
Little Gull	accidental	BARN OWLS	
Common Black-headed Gull	accidental	Barn Owl	regular
Bonaparte's Gull	regular		
Mew Gull	accidental	TYPICAL OWLS	
Ring-billed Gull	regular	Eastern Screech-Owl	regular
California Gull	accidental	Great Horned Owl	regular
Herring Gull	regular	Snowy Owl	regular
Thayer's Gull	casual	Northern Hawk Owl	accidental
Lesser Black-backed Gull	casual	Burrowing Owl	casual
Slaty-backed Gull	accidental	Barred Owl	regular
Glaucous Gull	regular	Great Gray Owl	accidental
Great Black-backed Gull	casual	Long-eared Owl	regular
Black-legged Kittiwake	casual	Short-eared Owl	regular
Sabine's Gull	accidental	Northern Saw-whet Owl	regular

Species	Status
GOATSUCKERS	
Common Nighthawk	regular
Chuck-will's-widow	regular
Whip-poor-will	regular
SWIFTS	
Chimney Swift	regular
HUMMINGBIRDS	
Ruby-throated Hummingbird	regular
Rufous Hummingbird	accidental
KINGFISHERS	
Belted Kingfisher	regular
WOODPECKERS AND ALLIES	
Lewis's Woodpecker	accidental
Red-headed Woodpecker	regular
Red-bellied Woodpecker	regular
Yellow-bellied Sapsucker	regular
Downy Woodpecker	regular
Hairy Woodpecker	regular
Three-toed Woodpecker	accidental
Black-backed Woodpecker	accidental
Northern Flicker	regular
Pileated Woodpecker	regular
TYRANT FLYCATCHERS	
Olive-sided Flycatcher	regular
Western Wood-Pewee	accidental
Eastern Wood-Pewee	regular
Yellow-bellied Flycatcher	regular
Acadian Flycatcher	regular
Alder Flycatcher	regular
Willow Flycatcher	regular
Least Flycatcher	regular
Eastern Phoebe	regular
Say's Phoebe	accidental

Species	Status
Vermilion Flycatcher	accidental
Great Crested Flycatcher	regular
Western Kingbird	regular
Eastern Kingbird	regular
Scissor-tailed Flycatcher	regular
LARKS	
Horned Lark	regular
SWALLOWS	
Purple Martin	regular
Tree Swallow	regular
Northern Rough-winged Swallow	regular
Bank Swallow	regular
Cliff Swallow	regular
Barn Swallow	regular
JAYS, MAGPIES, AND CROWS	
Gray Jay	accidental
Blue Jay	regular
Pinyon Jay	accidental
Clark's Nutcracker	accidental
Black-billed Magpie	accidental
American Crow	regular
Common Raven	accidental
TITMICE	
Black-capped Chickadee	regular
Boreal Chickadee	accidental
Tufted Titmouse	regular
NUTHATCHES	
Red-breasted Nuthatch	regular
White-breasted Nuthatch	regular
Pygmy Nuthatch	accidental

Species	Status	Species	Status
CREEPERS		**WAXWINGS**	
Brown Creeper	regular	Bohemian Waxwing	regular
		Cedar Waxwing	regular
WRENS			
Rock Wren	accidental	**SHRIKES**	
Carolina Wren	regular	Northern Shrike	regular
Bewick's Wren	casual	Loggerhead Shrike	regular
House Wren	regular		
Winter Wren	regular	**STARLINGS AND ALLIES**	
Sedge Wren	regular	European Starling	regular
Marsh Wren	regular		
		VIREOS	
OLD WORLD WARBLERS, OLD WORLD		White-eyed Vireo	regular
FLYCATCHERS, THRUSHES, AND WRENTITS		Bell's Vireo	regular
Golden-crowned Kinglet	regular	Solitary Vireo	regular
Ruby-crowned Kinglet	regular	Yellow-throated Vireo	regular
Blue-gray Gnatcatcher	regular	Warbling Vireo	regular
Eastern Bluebird	regular	Philadelphia Vireo	regular
Mountain Bluebird	accidental	Red-eyed Vireo	regular
Townsend's Solitaire	casual		
Veery	regular	**WOOD WARBLERS, BANANAQUITS, TANAGERS,**	
Gray-cheeked Thrush	regular	**CARDINALS, GROSBEAKS, EMBERIZINES,**	
Swainson's Thrush	regular	**BLACKBIRDS, AND ALLIES**	
Hermit Thrush	regular	Blue-winged Warbler	regular
Wood Thrush	regular	Golden-winged Warbler	regular
American Robin	regular	Tennessee Warbler	regular
Varied Thrush	regular	Orange-crowned Warbler	regular
		Nashville Warbler	regular
MOCKINGBIRDS, THRASHERS, AND ALLIES		Northern Parula	regular
Gray Catbird	regular	Yellow Warbler	regular
Northern Mockingbird	regular	Chestnut-sided Warbler	regular
Sage Thrasher	accidental	Magnolia Warbler	regular
Brown Thrasher	regular	Cape May Warbler	regular
Curve-billed Thrasher	accidental	Black-throated Blue Warbler	regular
		Yellow-rumped Warbler	regular
WAGTAILS AND PIPITS		Black-throated Gray Warbler	accidental
American Pipit	regular	Townsend's Warbler	accidental
Sprague's Pipit	accidental	Black-throated Green Warbler	regular
		Blackburnian Warbler	regular

Species	Status	Species	Status
Yellow-throated Warbler	regular	Vesper Sparrow	regular
Pine Warbler	regular	Lark Sparrow	regular
Prairie Warbler	regular	Lark Bunting	casual
Palm Warbler	regular	Savannah Sparrow	regular
Bay-breasted Warbler	regular	Grasshopper Sparrow	regular
Blackpoll Warbler	regular	Henslow's Sparrow	regular
Cerulean Warbler	regular	Le Conte's Sparrow	regular
Black-and-white Warbler	regular	Sharp-tailed Sparrow	regular
American Redstart	regular	Fox Sparrow	regular
Prothonotary Warbler	regular	Song Sparrow	regular
Worm-eating Warbler	regular	Lincoln's Sparrow	regular
Ovenbird	regular	Swamp Sparrow	regular
Northern Waterthrush	regular	White-throated Sparrow	regular
Louisiana Waterthrush	regular	Golden-crowned Sparrow	accidental
Kentucky Warbler	regular	White-crowned Sparrow	regular
Connecticut Warbler	regular	Harris's Sparrow	regular
Mourning Warbler	regular	Dark-eyed Junco	regular
Common Yellowthroat	regular	Lapland Longspur	regular
Hooded Warbler	regular	Smith's Longspur	regular
Wilson's Warbler	regular	Chestnut-collared Longspur	accidental
Canada Warbler	regular	Snow Bunting	regular
Yellow-breasted Chat	regular	Bobolink	regular
Summer Tanager	regular	Red-winged Blackbird	regular
Scarlet Tanager	regular	Eastern Meadowlark	regular
Western Tanager	accidental	Western Meadowlark	regular
Northern Cardinal	regular	Yellow-headed Blackbird	regular
Rose-breasted Grosbeak	regular	Rusty Blackbird	regular
Black-headed Grosbeak	accidental	Brewer's Blackbird	regular
Blue Grosbeak	regular	Great-tailed Grackle	regular
Lazuli Bunting	accidental	Common Grackle	regular
Indigo Bunting	regular	Brown-headed Cowbird	regular
Dickcissel	regular	Orchard Oriole	regular
Green-tailed Towhee	accidental	Northern Oriole	regular
Rufous-sided Towhee	regular		
American Tree Sparrow	regular		
Chipping Sparrow	regular	FRINGILLINE AND CARDUELINE FINCHES AND ALLIES	
Clay-colored Sparrow	regular	Rosy Finch	accidental
Field Sparrow	regular	Pine Grosbeak	casual

Species	Status	Species	Status
Purple Finch	regular	American Goldfinch	regular
House Finch	regular	Evening Grosbeak	regular
Red Crossbill	regular		
White-winged Crossbill	regular	OLD WORLD SPARROWS	
Common Redpoll	regular	House Sparrow	regular
Hoary Redpoll	accidental	Eurasian Tree Sparrow	casual
Pine Siskin	regular		

Index to Species Accounts

Bur Oak Books

A Cook's Tour of Iowa
By Susan Puckett

The Folks
By Ruth Suckow

Fragile Giants: A Natural History of the Loess Hills
By Cornelia F. Mutel

An Iowa Album: A Photographic History, 1860–1920
By Mary Bennett

Iowa Birdlife
By Gladys Black

Landforms of Iowa
By Jean C. Prior

More han Ola og han Per
By Peter J. Rosendahl

Neighboring on the Air: Cooking with the KMA Radio Homemakers
By Evelyn Birkby

Nineteenth-Century Home Architecture of Iowa City: A Silver Edition
By Margaret N. Keyes

Nothing to Do but Stay: My Pioneer Mother
By Carrie Young

Old Capitol: Portrait of an Iowa Landmark
By Margaret N. Keyes

Parsnips in the Snow: Talks with Midwestern Gardeners
By Jane Anne Staw and Mary Swander

A Place of Sense: Essays in Search of the Midwest
Edited by Michael Martone

Prairies, Forests, and Wetlands: The Restoration of Natural Landscape Communities in Iowa
By Janette R. Thompson

A Ruth Suckow Omnibus
By Ruth Suckow

"A Secret to Be Burried": The Diary and Life of Emily Hawley Gillespie, 1858–1888
By Judy Nolte Lensink

Tales of an Old Horsetrader: The First Hundred Years
By Leroy Judson Daniels

The Tattooed Countess
By Carl Van Vechten

"This State of Wonders": The Letters of an Iowa Frontier Family, 1858–1861
Edited by John Kent Folmar

Townships
Edited by Michael Martone

Vandemark's Folly
By Herbert Quick

The Wedding Dress: Stories from the Dakota Plains
By Carrie Young